C0-APN-479

Purification of Wilderness Waters

—A Practical Guide—

David O. Cooney, Ph.D.

Balsam Books
Laramie, WY 82072

© 1998 David O. Cooney

All rights reserved.
This book may not be duplicated in any way without the express written consent of the publisher, except in the form of brief excerpts or quotations for the purposes of review. The information contained herein is for the personal use of the reader and may not be incorporated in any commercial programs or other books, databases, or any other kind of software without the written consent of the publisher. Making copies of this book, or any portion of it, for any purpose other than your own, is a violation of United States copyright laws.

Balsam Books
1070 Inca Dr.
Laramie, WY 82072

PRINTED IN THE UNITED STATES OF AMERICA

Book Production by Phelps & Associates, Lancaster, Ohio

Publisher's Cataloging-in-Publication
(Provided by Quality Books, Inc.)

Cooney, David O.
   Purification of wilderness waters : a practical
guide / David O. Cooney. -- 1st. ed.
   p. cm.
   LCCN: 99-75562
   ISBN: 0-9675854-1-4

   1. Water--Purification. 2. Wilderness survival.
I. Title.

TD433.C66 2000   628.1'6
                QBI99-1517

CONTENTS

PREFACE

Increased use of our wilderness areas for hiking, backpacking, kayaking, rafting, fishing, hunting, and other activities, has resulted in a greater incidence of ailments caused by drinking untreated water from streams, rivers, and lakes.

The most common disease picked up from backcountry waters is giardiasis, caused by the protozoa *Giardia lamblia*. A recent article by Mark Jenkins ("What's in the Water?", Backpacker Magazine, December 1996, pp.56-60) mentions a study of streams in Colorado by Dr. C. Hibler of Colorado State University. Hibler stated: "We didn't find a single stream, not one, without *Giardia*." It is not clear if the presence of *Giardia* is a recent phenomenon, or whether it has been around for a long time. One news article claims that *Giardia* have been in wild waters for more than 300 years. One of the supposed primary transmitters of the infection, beavers, used to be present in large numbers in wilderness areas of the USA and Canada. It seems extremely likely that explorers, trappers, and "mountain men" must certainly have picked up the disease over the past couple of centuries. And yet, one does not hear of this. Perhaps these people built up an immunity after having the disease one or two times.

In any event, we now recognize such diseases fairly quickly, when they occur, and have become concerned about how to avoid them. This book is an overview of the hazards contained in wilderness waters, mainly the biological ones (microorganisms), and a summary of various ways of removing and/or inactivating such organisms. The book describes much about the physical and chemical principles underlying the various methods, so that readers may gain a greater understanding of why various purification methods work or fail to work.

The writer has tried to be as exact and quantitative as possible when discussing things like filter pore sizes, disinfectant doses, disinfectant contact times, etc. However, methods like filtration and disinfection are not exact—they depend first of all on the particular type and strain of organism involved. The efficacies of the various methods also depend strongly on the condition of the water, especially its temperature, pH (an index of the acidity of the water), and whether or not it contains organic matter. Thus, there can be no exact set of rules. The guidelines presented in this book are approximate and, hopefully, conservative. They are the author's best estimate of proper procedure. However, the author accepts no responsibility for the failure of any of the methods to prevent illness. It is the obligation of each reader to take the information presented, and blend it with one's own experience and good judgement, in making personal decisions about water treatment.

The author invites comments from readers about anything in this volume so that, if future editions are published, the contents can be continually updated to make them more complete and more accurate.

One word about the term "wilderness waters." There are many areas carrying official U.S. government designation as "wilderness." The term "wilderness" in this book is, of course, not meant to be limited to only these areas. The term is meant broadly, to refer to any area where water must be taken from lakes, streams, and rivers without purification by municipal water treatment methods. The author considered using the term "back country," among others, but opted for "wilderness."

TERMINOLOGY

This is a good place to discuss the scientific terminology that we will use in this book.

• *Volumes* will be given in quarts, gallons (4 quarts), liters (L), and milliliters (mL). Obviously, there are 1000 mL in a liter. Since a quart is almost a liter, 0.946 liters to be more precise, we will often refer to them interchangeably. Sometimes we will refer to fluid ounces. There are 32 fluid ounces in a quart (1 fluid ounce equals 1 ounce of weight), and thus a fluid ounce is 946 mL/32 = 29.6 mL. This is usually is rounded off to 30 mL when solutions (e.g., tincture of iodine) are packaged; that is, the package label will say "1 fluid ounce, 30 mL."

• **Weights** will usually be quoted in grams (g), milligrams (mg), or pounds. One pound equals 454 grams.

• **Concentrations** of things like iodine in water will be given in mg/L or ppm. The symbol ppm stands for parts per million, by weight. Thus, an iodine concentration of 4 ppm means 4 grams of iodine per million grams of water, 4 mg iodine per million mg water, etc. Since one liter of water is almost exactly 1000 g (the density of water at normal temperatures is about 1.0 g/mL), a liter of water is about 1 million mg. This means that concentrations in mg/L and in ppm are essentially identical, and indeed we will use them interchangeably in this book.

• **Temperatures** will appear both in degrees Celsius (°C), which used to be called degrees Centigrade, and in degrees Fahrenheit (°F). The relation between the two is: $°F = (9/5)°C + 32$, and $°C = (5/9)(°F-32)$. The melting point of ice is 0°C or 32 °F, while water's normal boiling point (at sea level) is 100 °C or 212 °F. The table below shows some specific equivalences between the two temperature scales.

Fahrenheit	Celsius
32	0
41	5
50	10
59	15
68	20
77	25
86	30
95	35

When possible, we will give preference to Fahrenheit over Celsius in the text, since readers will probably be more familiar with the Fahrenheit scale, or else cite the temperature in both scales, using a format of the type "59 °F (15 °C)." When we use this latter approach, we will select pairs of values which are integers (i.e., numbers without decimals) in both scales.

Several "standard" temperatures are used in science and engineering. One is "room" temperature, 68 °F or 20 °C, and another is standard atmospheric temperature, 59 °F or 15 °C. In later discussions, where we must specify temperatures to illustrate certain points, we will usually select one of these two.

In reporting the results of scientific studies, Celsius will be used almost exclusively, since scientists work in that scale.

• **_The parameter pH_** will be extremely important to us, and will appear often. Pure water, H_2O, may be visualized as HOH. However, not all of the water molecules are in this form. A few split according to HOH $\rightarrow$ H^+ + OH^- to give "acidic" hydrogen ions H^+ and "basic" hydroxyl ions OH^-. The pH of a solution is an indicator of the concentration of H^+ ions. Neutral water has 0.0000001 moles of H^+ per liter, an awkwardly small number (a "mole" is just the number of grams of a substance equivalent to its molecular weight; thus one mole of elemental iodine, I_2, molecular weight 253.8, is 253.8 grams, while for H^+, molecular weight 1.0, 1 mole is 1 gram). pH is defined as the negative of the common logarithm of the H^+ concentration in moles/liter. With this definition, the pH of neutral water becomes 7, a much more manageable number. Acidic waters have pH values lower than 7 (each drop of 1 unit on the pH scale represents a 10-fold increase in the acid ions H^+), while alkaline waters have pH values greater than 7 (each increase of 1 pH unit represents a 10-fold decrease in H^+). pH has a crucial role in determining the forms of dissolved substances like iodine and chlorine, some of which have much more disinfecting power than others, and in affecting the speed of chemical reactions.

With this background in terminology, let us now begin our survey of how to purify wilderness waters.

1

Chemical and Biological Hazards

Contaminants that might be encountered in wilderness waters fall into two classes—chemical contaminants, and pathogens.

A. Chemical Contaminants

Possible chemical pollutants include inorganic substances such as arsenic, heavy metals, and fertilizers (nitrates, phosphates), and organic substances such as herbicides and pesticides.

Fortunately, most wilderness waters do not contain much in the way of inorganic contaminants, herbicides, and pesticides. Some waters might contain decaying organic matter (leaves and other vegetation), but while this may be unpleasant in terms of taste and appearance, it may pose no health hazard. Of the various ways to treat wilderness waters, none (with the exception of some expensive and heavy reverse osmosis systems which are used on seagoing vessels to desalinate water) can remove inorganic substances. As for organic substances, boiling can ultimately drive them off if they are volatile, but otherwise only adsorption by activated carbon filters will work (see later discussion of this). This book will not discuss the removal of chemical contaminants in any detail, but instead will focus on the removal or inactivation of biological contaminants.

B. Biological Contaminants

Pathogens (i.e., disease causing microorganisms) can be classified as protozoa, bacteria, and viruses (the class of organisms called Rickettsia fall somewhere between bacteria and viruses, but they are not water-borne, so we will ignore them).

1. Protozoa

Protozoa are single-celled free-living animal-like organisms which are very large compared to bacteria and viruses. Nearly all protozoa exist in the aquatic environment. Pathogenic protozoa account for about 10,000 of the 35,000 known species, and cause some very nasty diseases. Parasitic protozoa (i.e., those that live in or off of a host) that are found in USA and Canadian waters include species of *Giardia*, the most common of which is *Giardia lamblia*, *Cryptosporidium parvum*, and *Entamoeba histolytica*. Indeed, *Giardia lamblia* and/or *Cryptosporidium parvum* have been detected in 90% of the surface waters of the USA. Unfortunately, protozoa are more resistant to disinfection by iodine and chlorine than are bacteria and viruses (in fact, *Cryptosporidium* cysts are not killed by either iodine or chlorine); however, being relatively large (at least for microorganisms), they can be easily filtered.

a. Giardia

Giardiasis is one of the most common waterborne diseases in the USA. It has occurred not only in western states (Colorado, Montana) but in the Eastern US (Pennsylvania, Massachusetts). Some outbreaks have involved large numbers of individuals.

The life cycle of the organism involves two forms—a cyst, in which the organism is encased in a hard shell, and the trophozoite form that leaves the cyst shell when conditions are favorable. The cysts are egg-shaped, being approximately 6 microns wide by 10 microns long. A micron, or micrometer, is one-millionth of a meter, or one-thousandth of a millimeter, and is usually given the symbol μ or μm. Only things of 100 microns or more can be seen by the naked eye. All water filters sold for wilderness use can easily remove *Giardia* cysts, even though they are flexible to a degree and can narrow down to 4-5 microns in diameter.

The hard shell which encases the cyst is about 0.3-0.5 microns thick and protects the organism from harsh environmental conditions. Surface

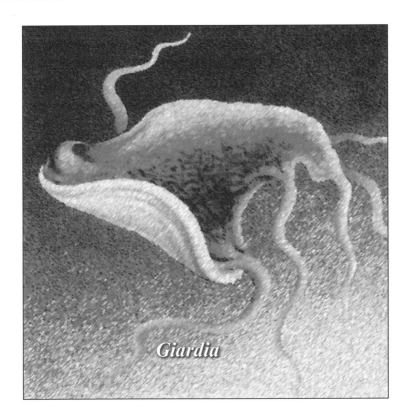

Giardia

waters contain the cyst form, usually deposited from the feces of infected animals. Researchers at Colorado State University have identified more than 30 species of animals as carriers of *Giardia*, among them beavers, dogs, muskrats, marmots, ground squirrels, cats, dogs, cattle, coyotes, and deer. Additionally, it has been estimated that from 3-20% of the U.S. population are carriers of the organism.

When a person drinks cyst-laden water, the trophozoites are released from their shells by the action of stomach acid and digestive enzymes. They attach themselves to the intestinal walls (usually in the upper small intestine) by means of a disk-shaped suction pad on their abdomens. Here they feed and reproduce, producing enormous numbers of offspring. The trophozoites are 10-20 microns long. Reproduction is by binary fission about every 12 hours, which means that a single trophozoite could produce a population of 1 million in 10 days, and 1 billion in 15 days. These large numbers of trophozoites cause acute irritation of the intestinal wall and block its normal functions to a high degree.

While in theory only one cyst is required for infection, studies have shown that roughly 10 or more are needed. Symptoms of the disease include severe diarrhea, abdominal cramps, flatulence, bloating, fatigue, nausea, foul stools, headaches, and weight loss. The incubation time is roughly 1-3 weeks. Left untreated, the disease will linger for several weeks or even months.

At some point in time, the trophozoite detaches from the intestinal wall and passes into the fecal stream in the lower intestines. Here, it transforms itself into the cyst form, and is excreted with the feces. The cyst form can subsequently infect other people or animals.

If treated with medications, giardiasis can be cleared up in a few days. Two prescription drugs are available, metronidazole (Flagyl), and furazolidine (Furoxone). Their efficacies in curing patients are both around 90%. The disease generally does not produce any permanent damage.

Giardiasis, in the USA and Canada, occurs with highest frequency in mountainous areas where water is taken from clear running streams or rivers, and is chlorinated but not filtered. In the USA, the population at risk for this is approximately 20 million. Hikers and backpackers exhibit the next highest incidence.

Giardia cysts can survive for a long time in water, especially if the water is cold. Under winter conditions (water temperatures 0-3 °C, 32-34 °F), *Giardia* cysts have been found to survive in lake and river waters for as long as 84 days.

The amount of chlorine that is usually used for disinfection of municipal water supplies is not generally effective in killing *Giardia*. Laboratory studies have shown that chlorine loses much of its disinfecting power when: (1) the water pH is above 7.5, (2) the water is cold, (3) the water contains significant organic content such as mud and decayed vegetation, (4) the contact time of the chlorine with the cysts is relatively short, and (5) the chlorine concentration is low (less than, say, 0.5 mg/L of free chlorine). These factors all work together, and one can compensate for one problem factor by improving another. Thus, cold water with a high organic content might be adequately disinfected if a higher than normal chlorine dose is applied for a longer than normal time (e.g., a dose of 5 mg/L for several hours). Conversely, organic-free water at pH

7 and 77 °F (25°C) might be adequately treated with 0.5 mg/L chlorine and a contact time of only 15 minutes.

For municipal water purposes, the best way to prevent giardiasis is to use filtration to physically remove the cysts. However, the conventional sand filters found in most water treatment plants are too porous to strain out such small cysts, and thus one must use flocculating or coagulating agents to cause the cysts and suspended matter to bind together into particles of filterable size. Chlorine is then added to kill bacteria and viruses which are not captured by the filter.

Giardiasis also occurs in Africa, Asia, and Latin America, with a total number of infections exceeding 2 million per year.

b. Cryptosporidium parvum ("Crypto")

This organism is a protozoan parasite whose cyst form is spherical with a diameter of 4-5 microns (thus, it is smaller than the cysts of *Giardia lamblia*). The cyst (more precisely called an oocyst) is, like that of *Giardia*, somewhat flexible, and can narrow down to about a 3 micron diameter. It is small enough to penetrate conventional rapid sand filters used in municipal water treatment plants. Upon ingestion, the cysts infect

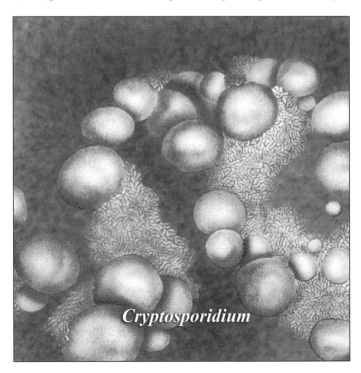

Cryptosporidium

the cells of the digestive tract, liver, kidneys, and blood. The entire life cycle takes place inside cells. The cysts incubate in 2 to 10 days, causing cholera-like illness: diarrhea, headache, abdominal cramps, nausea, vomiting, and low-grade fever. *Cryptosporidium* is found not only in lakes, rivers, and streams, but also in jacuzzis. Chlorine and iodine in usual doses do not kill the cyst form; however, boiling water for one minute does. Filters must be of about a two micron size or smaller to be sure of removing the cysts. *Cryptosporidium* organisms are found in humans and both wild and domesticated animals. Carriers may not show any evidence of the disease.

Symptoms usually last for about two weeks or less, but sometimes persist for several weeks. Relapses, more severe than the original bout, can occur after an apparent cure. Unlike the situation with giardiasis, there is no drug which can cure cryptosporidiosis.

This organism was first recognized as a cause of illness in 1976. Over the next 6 years, cryptosporidiosis was reported rarely, but in 1982 the number of cases rose sharply. Large outbreaks from municipal water sources have occurred in Texas (1984), Georgia (1987), Oregon (1992), and Nevada (1994). A huge outbreak involving more than 400,000 persons took place in Milwaukee in 1993 (30 died). A 1995 article in the Washington Post (June 16, 1995) claims that between 60,000 and 1.5 million Americans each year become ill from this organism.

Several studies have suggested that the number of crypto cysts required to cause disease may be somewhat larger than with *Giardia*, e.g. on the order of 100 cysts for a 20% chance of infection.

Unfortunately, *Cryptosporidium* cysts are very resistant to disinfectants. In one study, the survival of crypto cysts exposed for 18 hours to a variety of disinfectants (formaldehyde, ammonia, sodium hydroxide, cresylic acid, hypochlorite, benzalkonium chloride, and iodoform) was examined. Only 10% formaldehyde solution and 5% ammonia solution were effective in killing the cysts.

One research study looked at the effects of ozone and chlorine dioxide, as well as chlorine and monochloramine, on crypto cysts. Greater than 90% inactivation, as measured by infectivity of mice, was produced by: (1) 1 ppm ozone for 5 minutes, (2) 1.3 ppm chlorine dioxide for 1 hour, (3) 80 ppm chlorine or 80 ppm monochloramine for 90 minutes. Tests with *Giardia* cysts showed that *Giardia* cysts are killed 30 times more easily by ozone and 14 times more easily by chlorine dioxide than are crypto cysts. It appears that the only practical disinfectant for crypto is ozone.

c. Giardia and Cryptosporidium in USA Surface Waters

The occurrence of crypto cysts in surface waters of the USA has been examined by several investigators. In 1988, researchers looked at 107 surface water samples taken from 10 sources (rivers, streams, lakes, reservoirs, treated effluents, raw sewage) in six Western states. Most of the samples were from sources receiving sewage discharges, and these often had significant crypto cyst counts. However, water samples taken from two streams having no wastes discharged into them averaged 0.04 and 0.05 cysts per liter. Assuming that, say, 5 cysts would cause disease, one would have to drink 100-125 liters of these waters to receive an infectious dose.

In 1991, the same researchers examined 257 water samples from rivers, streams, lakes, reservoirs, and springs in 17 states (AZ, AR, CA, CO, CT, FL, GA, HI, MA, MI, MO, NY, OR, PA, TX, UT, WA) and found crypto cysts in 55% of the samples, at an average concentration of 0.43 cysts/liter. *Giardia* cysts were found in 16% of the same samples, at an average concentration of 0.03 cysts/liter. Samples from "pristine" sources (those coming from watersheds with little human activity, and receiving no agricultural or domestic sewage discharges) had crypto and *Giardia* cyst levels 11 and 13 times less, respectively, than waters from "polluted" sources. Crypto cysts were found in 39% of the pristine samples, at an average concentration of 0.2 cysts/liter, and *Giardia* cysts were found in only 7% of the pristine samples, at an average concentration of 0.004 cysts/liter. Assuming 5 cysts of either type would be needed to produce infection, one would have to drink 25 liters to get cryptosporidiosis and 1250 liters to get giardiasis.

In a 1991 study, waters were sampled by researchers in 13 Eastern and Midwestern states (CT, NJ, PA, MD, WV, VA, KY, TN, OH, IN, IL, MO, IA) plus one site in California (Bay area), and one site in a Canadian province (Alberta). *Giardia* and *Cryptosporidium* cysts were found in 81% and 87% of the waters, respectively (97% of the waters contained at least one of the two). The samples were all taken from waters being sent to water treatment facilities. Higher cyst counts were found to be associated with sources receiving industrial or sewage effluents, and cyst counts generally correlated with indices of poorer water quality (fecal coliform counts, turbidity). Water samples from "protected watersheds" contained a maximum of about 0.3 *Giardia* cysts/liter and 4 crypto cysts/liter. Since the states involved in the survey are generally of high population density, and industrialized or heavily agricultural, these data say little about waters taken from more pristine sources.

In 1987, researchers sampled four rivers in Washington State and two rivers in California, and found crypto cysts in all samples, with a range of concentration of 2-112 cysts/liter (average value = 25 cysts/liter). However, many of the sampling sites were downstream of large dairy farming areas, and cattle in many of those areas are known to carry crypto. The two California rivers were in the middle of a large urban area and downstream of major agricultural and livestock areas. Thus, the relevance of this study to backcountry waters is highly questionable.

In a 1989 study, relatively pristine rivers in Washington State were sampled for *Giardia* cysts, and cysts were found in 43% of 222 samples, with an overall average concentration of 0.063 cysts/liter. Average concentrations in the Cedar, Green, and Tolt rivers were 0.17, 0.20, and 0.13 cysts/liter, respectively, for those samples that did contain cysts. These values are relatively low. A dose of 5 cysts would require drinking 25-38 liters.

d. Other Protozoa

Entamoeba histolytica is a protozoan that infects mainly humans and other primates. Its cysts are about 10 microns in size. Various animals can become infected, but do not shed cysts, and thus are not involved in transmission of the disease. Symptoms and onset time of the disease, which is called amebiasis, or amebic dysentery, are highly variable, depending on the particular strain of the organism involved. The main effects of this organism in man are dysentery and ulceration of the colon and liver. The most dramatic incidence of infection in the USA was at the Chicago World's Fair in 1933, a result of contaminated drinking water due to defective plumbing in a sewage system. There were 1000 cases, with 58 deaths. In recent times, isolated cases due to transmission by food handlers have occurred, but no single large outbreaks have occurred. This organism is presently not a hazard in wilderness waters of the USA and Canada. Treatment with the drugs metronidazole and dilozanide furoate is effective.

Cyclospora cayetanensis, once thought to be linked to blue-green algae, has been identified as a new protozoan parasite, related to *Cryptosporidium*. Its cyst is spherical and 8-10 microns in diameter. Symptoms of the disease develop in about 4-7 days, last from 1-4 weeks, and are very similar to those of cryptosporidiosis. This organism had never been seen in humans before 1977, but became increasingly linked to "traveler's diarrhea" in persons returning from Mexico and Haiti in 1986 and later. About 20 people in Chicago were infected from contaminated water in 1990. In 1996, outbreaks occurred in New York and Florida, and in 1997 over 1000 persons were infected in Florida and Texas. It has been suspected that the organism is carried on fresh fruit and produce, particularly raspberries (from Guatemala) and strawberries, but this has not been proved conclusively. Outbreaks in 1997 have implicated contaminated baby lettuce (cases in Florida in March and April) and basil (cases in Virginia and Maryland in June and July). One health official stated: "It looks like any fresh produce has the possibility of contamination" (*USA Today*, July 23, 1997). It is not clear whether contaminated water is the primary transmission agent of *Cyclospora* to the various types of produce. The disease can be treated with antibiotics (Bactrim, Septra).

2. Bacteria

Bacteria are next in size to protozoa, being on the order of 0.3-2 microns. They are unicellular, reproduce by fission, and come in various shapes: spherical (coccus), rod-like (bacillus), and spiral (spirillum). A filter having a pore size of 0.2 microns will remove them with a high degree of certainty. Iodine and boiling will kill bacteria, as will chlorine if used correctly.

Bacteria that are found in water include enteric species (those that are found in the intestinal tract) such as E. coli (*Escherichia coli*), *Shigella* species, and *Salmonella* species. In the USA, *E. coli, Campylobacter jejuni, Shigella sonnei, Shigella felxneri,* and *Salmonella* have been implicated in waterborne disease outbreaks. These bacteria range in size from about 0.3 microns (*Campylobacter*) to about 0.6 microns (*Salmonella*). They cause diseases variously characterized by diarrhea, fever, vomiting, and dysentery (a combination of pain, fever, and severe diarrhea). In under-developed countries, waters often harbor various typhoid and cholera (e.g., *Vibrio cholerae*, 0.5 microns) bacteria.

Bacteria, like protozoan cysts, can survive in cold water for several weeks. The "dose" of bacteria required to produce disease varies widely, from roughly 10 to about 1000. Fortunately, there are prescription drugs for treating essentially any type of bacterial infection (however, very recently, a strain of *Staphylococcus* was found in Japan which no known antibiotic can kill; the "last hope" antibiotic, vancomycin, is not effective against this new strain).

While some strains of *E. coli* are pathogenic, most are not. Indeed, *E. coli* are one species of bacteria which are needed in one's intestines for digestion and good health (in the absence of intestinal bacteria, we could not live). Thus, *E. coli* are present in sewage in large numbers, and thus their presence in water indicates contamination by sewage, which may contain other bacteria and microorganisms which <u>may</u> be harmful to persons drinking such waters, i.e., it is an "indicator" species.

Researchers have studied how well iodine kills a wide variety of bacteria. The bacteria included *E. coli*, *S. typhosa*, *S. flexneri*, *S. sonnei*, *S. dysenteriae*, and five others. Temperatures were 2-5 °C (about 36-41 °F) and 20-26 °C (68-79 °F), the contact times were 1, 2, or 5 minutes, and the pH levels were 6.5, 7.5, 8.5, and 9.15. They determined the iodine doses in ppm required for a 100% kill. Under the "worst" conditions of pH 9.15 and 2-5 °C, an average minimum iodine concentration of 4.3 ppm killed all of the bacteria in 1 minute. Under the "best" conditions of pH 6.5 and 20-26 °C, an average minimum iodine concentration of 0.6 ppm killed all of the bacteria in 1 minute. For a contact time of 5 minutes and the "worst" conditions, the iodine concentrations needed to kill the bacteria ranged from 0.4 to 1.8 ppm. We may conclude that, even for cold water and high pH, bacteria are killed relatively easily by iodine.

3. Viruses

Viruses are the smallest of the pathogens. They consist of a core of nucleic acid, surrounded by a coat of protein (some viruses also contain lipid and/or carbohydrate material). They are rod-shaped or spherical, and generally range in size from about 0.02 microns to 0.1

microns. Because of their extremely small size, no commercially available wilderness water filters can remove them (one filter, the First Need Deluxe, claims to remove viruses, but its 0.4 micron pore size makes this claim questionable).

The reverse osmosis systems used for seawater desalination on sea-going vessels can filter them, since viruses are larger than the pores in "RO" membranes, but such systems are not practical for backcountry use. Filters containing an iodine matrix can kill them, as will iodine and boiling.

Common viruses in water are hepatitis A, Norwalk viruses, rotaviruses, and echoviruses (polioviruses also exist, but most people have been vaccinated against them). Virus incubation times are about 2-3 days, and they can cause a wide range of harm. Viral infections are generally characterized by intestinal discomfort, headache, fever, diarrhea, and specific disease states such as hepatitis and polio.

There are no drugs which can cure viral infections. Fortunately, viruses are not much of a problem in wilderness areas of the USA, Canada, and Europe. However, they are a real threat in many other places (e.g., the Himalayas).

2

Initial Clarification

Depending on the amount of dirt and debris in the water, a preliminary clarification step to remove such physical impurities may be warranted. If this step appears to be needed, there are various ways of carrying it out. First, it is assumed that any debris will be removed by pouring off floating debris from the collection vessel, or pouring off the water above sunken debris into a second vessel.

A. Settling

Settling will greatly extend the life of any water filter that one intends to use. Letting the water sit for as long as practicable (e.g., 12-24 hours in a large container) will allow sediment to fall to the bottom. After settling is complete, one should carefully pour or siphon the water into a clean container.

Coarse sand (0.5-1 mm diameter) settles rapidly, at rates between 600-1200 ft/hr (2-4 inches/second), whereas fine sand (0.10-0.25 mm diameter) settles at about 90-300 ft/hr (0.3-1 inches/second). Silt (0.005-0.05 mm diameter) settles at rates between 0.5-0.35 ft/hour, and fine clay (0.0001-0.001 mm diameter) settles at rates from 0.0002-0.02 ft/hr. It is obvious that the settling of fine clay and colloidal matter may take considerable time, unless accelerated by a chemical coagulant/flocculent.

An effective coagulation/flocculation agent that can be used is alum (aluminum sulfate), the agent used most commonly in municipal water treatment facilities. It is available in some grocery stores as "pickling powder." The dose is 1/8 to 1/4 tsp per gallon. Alum is particularly effective for waters containing a lot of organic material, such as decayed vegetation. Alum is also the ingredient used for "styptic" pencils, employed to stem bleeding from small cuts (e.g., those occurring during shaving). Unless alum is allow to settle well from treated water, it imparts a horrible taste.

B. Coffee Filters

These are excellent for a preliminary filtering of particulate matter from water. Simply place one filter, or a stack of 2-4 (one inside the other), in the opening of a jar, and pour the dirty water into the filter. The process is often slow. With muddy water, it will produce about a quart of cleaned water in two hours. The filters must be replaced when they get sufficiently plugged up.

C. Can Filters

Any porous material such as cloth (e.g., a handkerchief or bandanna), paper towels, mats of dried grass, and granular materials such as crushed charcoal and sand all can be made to work as coarse filters. The dried grass or granular materials can be placed in a large metal can, if available, which has had small holes punched into the bottom (e.g., with a nail). A clean container should be placed beneath the can which contains the filtering material, so as to collect the filtered water.

The material employed must be packed tightly against the sides of the can in order to prevent water from flowing around it, thereby avoiding filtration. Depending on the filtering material used, its depth, and the can size, this method can produce up to two gallons per hour of clean water.

D. Capillary Siphoning

In this method, one elevates the container filled with dirty water above a second, clean container. Then, one runs a piece of yarn, strips of cloth (cotton works well), strips of a towel (terry-cloth is best), etc. from the top container to the bottom one. The chosen material should be soaked in clean water to get things started. The yarn/cloth/towel material will draw water through itself by capillary action, leaving dirt and debris behind. The drawback to this method is that it is very slow, producing only about a cup of clean water per hour.

E. Hose Siphoning

One takes a hose or tube of arbitrary size, and stuffs a ball of cotton (not too tightly) into one end. This end is placed in the dirty water container, which is elevated above a clean collection container. By sucking on the lower end of the hose or tube, one can initiate siphoning action. The cotton ball should be replaced when it starts to plug up to a significant extent. Depending on the diameter of the hose or tube, the height difference between the containers (increasing this gives a greater gravity driving-force), the dirtiness of the water, and the frequency of replacement of the cotton ball, one can produce as much as two quarts of clean water per hour with this method.

Subsequent to any clarification that may be carried out, here are three basic methods of purifying backcountry waters: (1) boiling, (2) filtration, and (3) disinfection with iodine, iodine compounds, or chlorine compounds. We will consider each of these in the following chapters.

3

Boiling

Boiling for 1-5 minutes will generally kill all protozoa, bacteria, and viruses that are found in wilderness waters. Protozoa are killed quickly by boiling, bacteria less quickly, and viruses more slowly. There are reports that some viruses found overseas can withstand boiling for up to 15 minutes.

Researchers have determined the "thermal death point" (TDP) of many protozoan cysts in pure water, which is the temperature at which 100% mortality will occur in 5 minutes. The TDP for *E. histolytica* was 68 °C (154 °F) and that for *Giardia intestinalis* was 64 °C (147 °F). For four other kinds of protozoan cysts the TDP values ranged from 64-76 °C (147-169 °F). Thus, a temperature of about 170 °F for 5 minutes should guarantee the destruction of protozoan cysts of virtually any kind.

At sea level, water boils at 212 °F (100 °C) and at higher elevations at somewhat lower temperatures. Table 3.1 shows the boiling points of water at different elevations.

The decrease in temperature with height is quite steady, with a fall of 1.83 °F (1.02 °C) per thousand feet. This agrees with a common "rule-of-thumb" of a 2 °F decrease per 1000 feet.

Since most organisms die very rapidly above 70 °C (158 °F), the table shows that boiling water at an elevation of 14,000 ft gives a temperature which is still much more than adequate for killing organisms. If

Table 3.1 **Boiling Point of Water at Different Elevations**

Elevation Above Sea Level (ft)	Boiling Point (°C)	Boiling Point (°F)
0	100.0	212.0
2000	98.0	208.3
4000	96.0	204.7
6000	93.9	201.1
8000	91.9	197.4
10000	89.8	193.7
12000	87.8	190.0
14000	85.7	186.3

one wishes to use a guideline of, say, one minute boiling at sea level, the boiling time should be increased significantly at higher elevations, since the mortality rate of organisms tends to decrease exponentially with a fall in temperature. One article in the literature recommends 5 minutes boiling at 10,000 ft, but this seems a bit excessive.

The flat taste of boiled water can be improved by aerating it, by pouring it back and forth into another container, although this entails the risk of recontamination if the other container or one's hands are not clean. Alternatively, one can add a pinch of salt, or add a drink mix such as lemonade, to improve the taste.

Boiling has some disadvantages: (1) it is relatively slow, as it takes several minutes for water to be brought from ambient temperature to boiling (boiling times for various backpacking stoves, with water initially at 70 °F, in still air at sea level, range from about 3.0 minutes to 5.3 minutes), (2) if one wants to drink the water soon after it has been boiled, one must wait a significant time for it to cool, (3) the consumption of valuable fuel, and the weight and cost it represents, is not trivial, if a fair amount of water is to be treated by boiling (if a camp fire is used to boil the water, there is the environmental insult of a fire-ring and ashes, not to mention the dirty pots which result), and (4) chemical pollutants having little or no volatility are not removed by boiling (e.g., inorganics such as arsenic and heavy metals; organics such as herbicides, pesticides, etc.). Fortunately, most wilderness waters do not contain much in the way of chemical contaminants. Some waters do contain decaying organic matter (leaves and other vegetation), which gives the water an unpleasant taste, odor, and appearance. Although such organic matter may pose no health threat, it may require using a higher dose of disinfectant, such as iodine or chlorine, because of binding of the disinfectant with the organic matter, leaving less "free" or "residual" disinfectant.

4

Filters and Filtration

Filters are quicker than boiling or chemical disinfection, and produce clean-tasting water (no chemical taste). Some filters also contain activated carbon to adsorb organic chemicals from the water, if they are present. However, filters are often heavy, much more so than a bottle of chemical disinfectant tablets or solution, and many require considerable force to operate their pumps. Their costs per gallon of water treated can approach and even exceed $1.00. However, the best devices produce pathogen-free water with a high degree of certainty, whereas with chemical disinfectants one is sometimes unsure if disinfection is complete.

A. "Filters" Versus "Purifiers"

Before discussing "filters" per se, we should mention the term "purifier" and distinguish it from "filter." Purifiers are defined by the EPA (Environmental Protection Agency) as units which "remove, kill, or inactivate all types of disease-causing microorganisms from the water, including bacteria, viruses, and protozoa cysts, so as to render the processed water safe for drinking" (US EPA Guide Standard & Protocol for Testing Microbiological Water Purifiers, 1987). Purifiers typically combine chemical disinfection with filtration.

Purifiers meeting EPA standards must remove 99.9999% of waterborne bacteria, 99.9% of protozoa, and inactivate 99.99% of viruses. The waters tested must include "worst case" types of waters—cold and turbid.

The EPA specifies that certain organisms be used in their test protocols: *Cryptosporidium* for protozoa (since it is smaller than *Giardia*, which was formerly the test organism), *klebsiella* for bacteria (this is not pathogenic, but it is easy to grow and handle), and poliovirus and rotavirus for viruses. The test protocols are rather involved, and will not be described here.

Most "purifiers" have a resin which releases iodine as water passes through, killing bacteria and viruses reasonably well, and some devices have either a silver-impregnated activated carbon bed or a silver-impregnated ceramic filter element. Silver is "bacteriostatic" and can inhibit the growth of bacteria; however, it does not act fast enough to kill bacteria in water passing through.

Resins have been prepared which contain triiodide (I_3) and pentaiodide (I_5) molecules. These resins are called Triocide and Pentacide, respectively, and can be purchased from Water Technologies Corporation (Ann Arbor, MI). They do release I_2 into water, but only in small concentrations (0.1-0.2 ppm). The way they kill bacteria is as follows: When bacteria physically contact the resin, I_2 transfers to the bacteria surfaces, then penetrates into the bacteria, and "devitalizes" them by inhibiting vital enzyme reactions in the bacteria.

Both the triiodide and pentaiodide resins are chemically and physically stable for long periods of time—as long as 15 and 5 years, respectively. The triiodide resin has been used extensively in the U. S. Space Program to ensure disinfection of water supplies.

Unlike the situation with purifiers, there are currently no federal standards for the performance of filters, but there are now major efforts to establish such standards through the ASTM (American Society for Testing and Materials) process.

B. Pore Size Considerations

There are a very wide variety of filters and purifiers on the market, and they have become quite popular. The filter "elements" have pores which cause particles and microorganisms of sizes larger than the pore size to be retained. Those of smaller size pass through. All available water filters and purifiers will remove larger organisms like protozoa, and many will remove bacteria. With one possible exception (the First Need Deluxe filter, described below), no available filters have pores small enough to filter out viruses.

Filters

Katadyn Pocket Filter

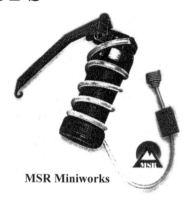

MSR Miniworks

MSR Waterworks II

PUR Pioneer

PUR Hiker

SweetWater Guardian

Pore sizes for filtration media (there are a very wide variety of such media used in all types of industries) are often given as "nominal" or "average." Both of these terms mean that there can be significant numbers of pores which are both larger and smaller than the stated size. The designation "absolute", in contrast, means that there are <u>no</u> pores larger than the size stated. Since some diseases can occur from ingesting just a few microorganisms, the absolute pore size rating is the only one that matters for water purification.

The pore sizes required to filter out various microorganisms are given in Table 4.1.

Table 4.1 **Pore Sizes Needed to Filter Specific Organisms**

Organism	Required Pore Size (microns)
Parasite eggs/larva	20
Giardia	5
Cryptosporidium	2
Bacteria	0.4
Viruses	0.01

There is a bacterium called *Brevundimonas diminuta* which is 0.3 microns in size, and is said to be equal in size to the smallest disease-causing bacteria. Thus, if we take 0.3 microns as the pore-size needed to remove all pathogenic bacteria with certainty, then filters of 0.5-1 micron pore-size could be marginal for bacteria removal, depending on what is in the particular water being treated. Many filters have pores in this size range.

C. Some Popular Filters and Purifiers

Table 4.2 gives some data on several models of water filters. All claim to filter protozoa and bacteria. Table 4.3 presents similar information for various purifiers. These claim to inactivate viruses in addition to removing protozoa and bacteria.

The filter manufacturers are: General Ecology (First Need), Exton, PA; Basic Designs, Santa Rosa, CA; Katadyn USA, Scottsdale, AZ; Mountain Safety Research, Seattle, WA; Recovery Engineering (PUR), Minneapolis, MN; SweetWater, Inc., Longmont, CO.; and Timberline Filters, Inc., Boulder, CO.

Table 4.2   **Popular Water Filters**

Name	Element Type	Pore Size (microns)	Weight (oz)	Approx. Price
First Need Deluxe	"structured matrix" microstrainer	0.4	15.0	$70
Basic Designs 2050 Ceramic Filter Pump	ceramic	0.9	8.0	$15
Katadyn Pocket Filter	ceramic	0.2	22.7	$295
Katadyn Minifilter	ceramic	0.2	8.0	$140
MSR MiniWorks	ceramic	0.3	14.3	$65
MSR Waterworks II	ceramic/ carbon	0.3	16.6	$140
PUR Hiker	pleated glass fiber/carbon core	0.5	11.0	$55
SweetWater Guardian	labyrinth depth filter/ carbon	0.2	11.0	$60
Timberline Eagle	polyethylene fibers/micro fiberglass	1.0	8.0	$20

Table 4.3   **Popular Water Purifiers**

Name	Element Type	Pore Size (microns)	Weight (oz)	Approx. Price
PUR Explorer	iodine resin/ glassfiber filter	1.0	24.8	$130
PUR Voyageur	iodine resin/ pleated glass fiber filter	0.5	11.0	$70
PUR Scout	iodine resin/ pleated glass fiber filter	1.0	12.0	$80
SweetWater Guardian Plus	glass fiber matrix/carbon/ iodine resin	0.2	14.5	$80

The purifier manufacturers are: Recovery Engineering (PUR), Minneapolis, MN; and SweetWater, Inc., Longmont, CO.

Table 4.4 summarizes information on how much pumping force is needed with various units, output rates in liters per minute, and the number of pumping strokes required to produce a liter of water. The output values range over a factor of 2.3, from 0.60 to 1.39 liters/minute. The strokes/liter values also show a factor of 2.3 variation, from 43 to 100. In contrast, the pumping force values vary by a factor of 12.5, from 1.6 to 20.0 pounds.

An article in the March 1996 issue of *Backpacker* magazine (D. Getchell, "Pure and Simple: How to Make Sure Your Backcountry Water is as Safe as What You Drink at Home," Backpacker, March 1996, pp. 172-183) gives a more extensive and detailed listing of available water filters, as well as water purifiers.

The term "carbon" in Table 4.2 means that the unit contains activated carbon (in other applications, this is sometimes called activated "charcoal"), the purpose of which is to adsorb organic chemical impurities,

Purifiers

First Need® Deluxe

PUR Voyager

PUR Explorer

PUR Scout

such as pesticides, herbicides, etc. Carbon can also adsorb molecular iodine (I_2) and molecular chlorine (Cl_2). Quoted prices are typical values obtained by checking major suppliers such as REI, Campmor, etc. and making a judgement as to a suitable average price, when disparities existed (most prices were quite similar from catalog to catalog). Device weights were also averaged in the same way. Again, weights were usually similar from catalog to catalog.

Table 4.4 **Pumping Force, Output, and Strokes per Liter**

	Pump Force (pounds)	Output (liters/min)	Strokes per liter
Water Filters			
First Need Deluxe	6	1.3	45
Katadyn Pocket Filter	20	0.7	80
MSR Miniworks	9	0.6	100
MSR Waterworks II	12	0.8	75
PUR Hiker	8	1.2	50
PUR Pioneer	2	1.0	60
SweetWater Guardian	2	1.0	60
Water Purifiers			
PUR Explorer	5	1.4	45
PUR Voyageur	1.6	1.1	55
PUR Scout	1.6	1.0	60

Among other things, Tables 4.2 and 4.3 show that some of the filters are fairly heavy, weighing as much as a pound and a half (the average weight is about 14 oz). It must be noted, in addition, that many of these devices retain significant water inside their filter units after use. One ends up carrying this water, as well as the "dry" weight of the device itself. This extra weight could easily be half a pound.

One disadvantage of ceramic elements is that they can crack if they freeze while water is in them. Their advantage is that they last a long time, as data given later will show.

The First Need Deluxe is listed here as a "filter," even though its manufacturer claims that it also removes viruses, and thus should be classified as a "purifier." No proof is known to the writer to back up this claim. If its pore size is indeed 0.4 microns, this is well above the upper limit of virus sizes, and thus virus removal would seem to be uncertain. We include the First Need Deluxe here as a "filter," not just for this reason, but because we will shortly discuss an extensive field test of "filters," and the First Need Deluxe was one of those involved.

D. Other Filter Models

Basic Designs also markets a 2040 Ceramic Water Filter (0.9 micron pore size, 24 oz, $79) which uses gravity rather than a pump to produce the flow, which is about 0.5 liters/min. There is also a lighter (12 oz) and cheaper ($64) gravity-flow version called the 2030 Ceramic Water Filter, which is very slow—about 0.07 liters/minute output rate. Replacement filters for the 2040 and 2030 filters are about $45 each, or three times the cost of the replacement filter for the 2050 unit.

Katadyn also offers a filter called the Katadyn Combi, which also has the same type of ceramic element, with the same pore size (0.2 microns) and weight (23 oz) of the Katadyn Pocket filter. However, it also contains activated carbon for adsorbing organic chemicals. At $185, it is a bit cheaper than the Pocket Filter.

A smaller version of the Pocket Filter is the Katadyn MiniFilter. It weighs 8 oz, puts out 0.5 liter/minute, and costs $139 (replacement filter element $59). Its pump handle has a different design from that of the Pocket Filter.

Katadyn also markets the Katadyn Water Filter (0.2 micron pore-size, 23 oz, $250), which has a ceramic element, and produces about 1 liter/minute, with a very large claimed lifetime—15,000 gallons. The lifetime of the Katadyn Pocket Filter is stated to be "many years", and that of the Katadyn Minifilter is given as 1000 gallons (although one source states 2000 gallons).

Still other Katadyn models include an Expedition unit ($875), and gravity-fed models called the Drip Filter ($275) and the Syphon Filter ($100). These will not be described in any detail here.

The long life of the Katadyn ceramic filters needs explanation. When the ceramic element becomes clogged, one must disassemble the unit (usually fairly easy) and use a metal file (supplied) to scrape off the clogged outer layer of ceramic, exposing fresh ceramic. The Katadyn instructions typically state that this can be done 100 times before the ceramic element must be replaced (a supplied measuring device is used to determine when replacement must be made). Thus, a "lifetime" of 1000 gallons means that the unit will typically last 1000/100 = 10 gallons between cleanings.

The Katadyn Pocket Filter requires a large pumping force, more than some individuals might be able to apply stroke after stroke. The Katadyn and MSR filters all require a fair number of strokes to produce a liter of water (76-100) and this makes their outputs in liters per minute much lower than all of the other filters and purifiers.

A recent field test of the Katadyn Minifilter by a couple on a 4-day backpack in Colorado showed that the filter clogged right away because the inlet line was inadvertently placed near a rock covered with brown algae. Cleaning was performed with relative ease. However, over the four days, 3 more cleanings were required while filtering 6-8 quarts per day of water which had "no visible particulates." That is, cleaning was needed about every 6-8 quarts. Additionally, the average time to filter one quart was 5 minutes for the man, who found that the "relative effort was high, requiring a 'death grip' on the filter body." The man further stated that "it took my wife about 10 minutes to pump a quart, including rest periods." The pump force required for this unit has been cited as 13 pounds, which is quite high.

PUR has a simplified filter model called the Pioneer, containing a glass fiber disk (11 square inches in area) of 0.3 micron pore-size. It weighs only 8.4 oz, and is inexpensive ($30). It is intended for "short trips or occasional use." Its pumping mechanism is similar to that of the PUR Hiker and PUR Voyageur, described below.

SweetWater also makes a SweetWater Walkabout Filter (9 oz, $35, pore-size 0.2 microns), which sells for about $35 and claims to remove over 99.9999% of bacteria and protozoa. It produces 0.9 liters/min and has a life of 125 gallons. SweetWater sells an attachment called the SweetWater Silt Stopper II (1 oz, $10) which can be attached as a pre-filter to their units, or any other manufacturer's units, to screen out silt and other large particulate matter. A pack of three replacement filters is about $9.

The Timberline Eagle has polyethylene fibers and micro fiberglass filtration elements. However, its absolute pore size is relatively large—1.0 microns—and thus it can not remove bacteria. A larger and more expensive gravity-flow device called the Timberline Basecamp (12 oz, $50) filter also has 1 micron pores.

Other filters of lesser popularity will be mentioned, to make this section more complete. American Camper (Lenexa, KS) sells a 0.9 micron Ceramic Water Filter ($34) with a weight of 8 oz and an output rate of 0.5 liter/minute. It has a prefilter. Based on its pore size, bacteria removal is questionable.

Three purifiers made by Sigg (Switzerland) and distributed by Outbound Products (Hayward, CA) are the Pocket Travel Well, the Trecker Travel Well, and the Sigg Microlite System. All have 1.0 micron pores. The first two have a depth filter with iodine; the third has a "proprietary" element. In the same order mentioned, their weights are 2, 6, and 7 oz; their output rates are 0.5, 1, and 1.5 liters/minute; and their prices are $35, $65 (replacement element $32), and $56 (the first and third units do not have replacment elements—they are throwaways). All claim to remove organic chemicals, so they must contain activated carbon. The 2 oz Pocket Travel Well sounds intriguing.

A company called Relags USA, Inc.(Boulder, CO), markets a couple of ceramic element filters. One, the Adventure Filter ($75, replacement element $35), has 0.3 micron pores, weighs 10 oz, and produces 0.5 liters/minute. The other, the Travel Filter ($150, replacement element $75), has 0.5 micron pores, weighs 23 oz, puts out 0.75 liters/minute. A comment on these, found on the web, suggests that they are somewhat difficult to pump.

E. Filter Field Tests

An article in *Backpacker Magazine* (K. Hostetter, "The Water Filter Field Test", Backpacker, December 1996, pp. 62-70, 112-115) involved field-testing of seven water filters, by 51 people, in Big Bend National Park (Texas) and the Blue Ridge Mountains (North Carolina). The filters were: Basic Designs 2050 Ceramic Filter Pump, First Need Deluxe, Katadyn Pocket Filter, MSR Miniworks, PUR Hiker, SweetWater Guardian, and the Timberline Eagle. The field testers were asked to comment on their ease of use, rate of output, ease of maintenance, packability (weight and bulk, tendency to leak in one's pack), and durability. The testers were also asked to vote for their favorite, their second favorite, and their least favorite.

The votes for the favorite, second favorite, and least favorite filters are shown in Table 4.5.

Table 4.5 Testers' Votes for Seven Filters Tested by *Backpacker Magazine*

Filter	Favorite	Second Favorite	Least Favorite
PUR Hiker	27	2	0
SweetWater Guardian	10	14	4
MSR Miniworks	11	9	3
Timberline Eagle	2	8	2
Katadyn Pocket Filter	0	10	5
First Need Deluxe	0	4	5
Basic Designs 2050	0	0	25

F. Detailed Comments on the Field Test Filters

We will give a synopsis the testers' comments, as reported in the *Backpacker* article, plus some selected comments taken from a web site called "http://www.wed-dzine.com/gearaddict/reviews." This web site contains reviews of all kinds of backpacking gear, submitted by people who wished to relay their experiences. We will present comments for the filters in the order shown in Table 4.5, from most favorite to least favorite.

1. PUR Hiker

As one can see from the sketch of this unit (p. 20), the PUR Hiker has a palm-sized handle on its pumping mechanism, an outlet hose containing a bottle "adapter" which can be placed in any sufficiently wide-mouth receiving bottle, and an inlet hose having a float "collar" and a coarse filter at the end. The float, also found in many other filters, keeps the inlet filter above the bottom of the water source (lake, river, etc.) and reduces uptake of dirt and debris. This unit has a 126 square-inch "Anti-Clog" pleated filter inside it, which is stated to be "ten times larger than any other microfilter on the market." The filter is "guaranteed not to clog for one year." However, the filter lifetime clearly depends on how dirty one's water sources are and how often one uses the unit. The "1 year" figure must simply be an estimate for a "typical" user.

We will present comments on this (and on other devices, below) in the form of lists of "Positives", "Negatives", and "Comments."

Positives

- Ergonomic design: palm-shaped pump handle; striated housing makes it easy to grip
- Pumping is quick and easy: fairly low force (8.0 lb) needed, good output (1.24 liters/minute), few strokes/liter (48)
- Very "packable": low weight (8 oz), minimal loose parts, and handy storage sack
- Has an activated carbon core to adsorb organic chemicals.

Negatives

- Needs to be primed. However, this is easy to do—one just unscrews the unit and fills the cartridge with water (one web site reviewer said that the instructions did not tell him to do this, and thus he couldn't get the unit to work)
- The bottle adapter doesn't fit any particular bottle, and so it doesn't "free up" a hand. An ordinary outlet hose would suffice.

Positive Comments from the Web

"a great filter"

"the pleated filter really extends the life"

"easy to pump"

"best filter I've ever had"

"excellent—fast, easy, and compact"

"has never let me down"

"ruggedly built, yet light and compact"

Negative Comments from the Web

"some priming problems"

"can't scrub the filter, but it lasts so long you might not need to"

"not as easily field-repaired as other filters"

"clogged fast with glacier water"

"cleaning the filter cartridge by swishing it in clean water didn't work"

Summary

Overall, the PUR Hiker came in first in every category of the *Backpacker* field test. It was generally regarded as the easiest unit to pump, with a superior rate of output, very packable, and easy to maintain. Its only real negative was the need to prime it—not a difficult thing to do, however.

2. SweetWater Guardian

As one can see from the sketch of this unit, the SweetWater Guardian has a lever type handle on its pumping mechanism, and—like the PUR Hiker—an outlet hose containing a bottle adapter, and an inlet hose having a float collar and a coarse filter at the end. Also, like the PUR Hiker, it contains activated carbon for adsorbing organic chemicals. The unit has a pressure relief valve which opens when the filter is clogged enough to require cleaning.

Positives
• Easy to clean with a brush (supplied with the unit)
• Pumping is very easy: only 2.0 lb force needed
• The output rate is good (1 liter/minute)
• The pressure relief valve tells you when the unit needs to be cleaned.
• Very "packable": low weight (11 oz), but the pump lever does need to be disengaged first.
• Has activated carbon to adsorb organic chemicals.

Negatives
• Required lots of maintenance (frequent cleaning). However, cleaning is easy to do.
• Like the PUR Hiker, the bottle adapter doesn't fit any particular bottle, and so it doesn't "free up" a hand. An ordinary outlet hose would suffice.
• The pressure relief valve, on top of the pump head, opens and squirts you in the face when pumping starts to get hard.
• The silt stopper prefilter plugs easily.

Positive Comments from the Web
"virtually problem-free for two seasons"
"only takes 5 minutes to clean"
"pump is efficient"
"ease and volume of pumping,"
"clogged quickly at first, but quickly remedied with the brush."

Negative Comments from the Web
"plunger mechanism failed—not field serviceable"
"plunger broke on the first day of a two-week trip"
"clogs very quickly in any kind of water—Appalachian, Great Lakes, Rockies"

Summary
Overall, the SweetWater Guardian essentially tied for second place

(with the MSR MiniWorks) in the *Backpacker* field test. It was easy to pump, but did require adequate cleaning to maintain good output. The web comments about plunger failure suggest possible problems with the durability of the pumping mechanism.

3. MSR MiniWorks

This filter (see sketch), like the SweetWater Guardian, has a lever-action pumping mechanism, and activated carbon for organic chemical removal. The base of the unit (water outlet point) is designed to screw into a 1 liter Nalgene bottle or an MSR Dromedary bag. The inlet hose has a float collar, and a relatively small coarse filter at the end. The filter element is ceramic.

Positives

• Sound ergonomics: the unit is easy to hold; pumping is easy.
• Only occasional cleaning needed—in field tests, it never clogged completely like other filters. Easy to clean.
• Ceramic filter element has a long life.
• Long intake hose is a convenience.
• Very "packable": few loose parts; fits into storage sack easily.
• If connected to the correct receiver, it delivers the water conveniently and frees-up a hand.
• Has activated carbon to adsorb organic chemicals.

Negatives

• Since there is no output hose, it is difficult to direct the water into a receiving vessel if the unit is not connected to either a 1 liter Nalgene bottle or MSR Dromedary Bag.

Positive Comments from the Web

"easy to pump; long-lasting filter"
"good-sized handle; better than palm-sized ones"
"great—easy to repair, clean, and use"

Negative Comments from the Web

"hard to direct the water into a bottle if it isn't the type to be connected to the unit"

Summary

Overall, the MSR MiniWorks essentially tied for second place (with the SweetWater Guardian) in the *Backpacker* field test. It was easy to hold and pump, gave good output, and was easy to maintain.

The MSR's ceramic filter lasts longer than the depth filter in the SweetWater Guardian. Since replacement prices for the two are comparable, the MSR unit provides better value.

4. Timberline Eagle

This unit (see sketch) has an up-and-down pump mechanism connected to a filter module which is immersed into the water source. Thus, there is no input hose or prefilter. Its pore size (1 micron) is too large to remove bacteria (and certainly viruses).

Positives

•High output rate (because of its large pores)

•Fast, easy pumping.

•Simple, and light in weight.

Negatives

•Large pores don't remove bacteria.

•Pump is hard to grasp: it is thin and slick.

•Clogs quickly, and cartridges are not cleanable (they are "throwaways").

•Flimsy appearance, and poor packability due to its long, awkward shape.

•Tended to leak in one's pack.

•Filter unit tends to detach from the pump if you pump too quickly.

•There is no way to back-flush the filter in order to try to clean it.

Comments from the Web: None found at the site used.

Summary

If one wants to remove only protozoa, and is willing to carry replacement cartridges, this unit provides a cheap, light, easy to use, and high output alternative to other filters.

5. Katadyn Pocket Filter

This filter contains a silver-impregnated ceramic element (silver inhibits bacterial growth). It has a T-shaped pump handle, and an inlet hose with a float collar, and a small prefilter at the end. The water outlet is a small nozzle near the top.

Positives

•The ceramic filter element, with proper cleaning, lasts a long time.

•The unit is made mostly of metal and is tough.

Negatives

•Its price, $295!

- Metal construction makes it heavy (23 oz).
- Hard to pump (20 lb force needed). Pump action stiff and difficult. T-shaped handle too small and hard to hold.
- Slow output.
- No output hose. Requires aiming the output stream accurately into a collection container, yet two hands are needed for pumping. Can't easily position the collection bottle and pump at the same time.
- In field tests, cleaning helped a lot, but only briefly.

Positive Comments from the Web
"has yet to fail me"

"a quick scrub gets it back in service"

"quality is superb and performance unwavering"

"still on my original filter, even though I've pumped lots of water"

Negative Comments from the Web
"sturdily built, but heavy, slow, tricep-taxing, expensive"

"overkill—15,000 gallon life!"

Summary
Takes too much time and energy to pump. Is its longevity worth the very high cost?

6. First Need Deluxe

This filter claims to removes viruses, even though its pore size is given as 0.4 microns—way above the size of viruses. This suggests that at least part of its filter matrix is much, much tighter than 0.4 microns. Articles in outdoor magazines indicate that, in the laboratory, it does remove viruses. However, virus removal in the field has not been confirmed. This unit (see sketch) has a coarse filter and prefilter on the inlet line, a double-action pump (i.e., it pumps water on both the upstroke and downstroke) with a T-shaped handle, and a short outlet line connected to the bottom.

Positives
- The double action pump gives a good output rate.

Negatives
- In the field, it clogged a lot (due to its tight filter). Its performance decreased rapidly with time. Thus, it has high maintenance.
- Pump force required is relatively high.
- The design is ergonomically deficient—the unit is tiresome to hold.
- The intake hose has no float; thus, the end of the hose drops to the

bottom of the water and sucks in debris. The prefilter and main filter both clog quickly.

• The filter can't be cleaned (however, it can be back-flushed, which could help a lot). A replacement element costs about $33.

• The intake hose pops off and sprays you if you try to pump too fast.

Comments from the Web: None found at the site used.

Summary

The hard pumping and need to carry replacement cartridges are trade-offs for the claimed virus-removing ability of this unit.

7. Basic Designs 2050 Ceramic Filter Pump

This filter is very similar in design to the Timberline Eagle. It has slender pump casing connected from the bottom by a hose to a filter unit. The filter consists of a foam cylinder containing a ceramic element inside.

Positives

• Light weight (8 oz)

Negatives

• Its large pore size (0.9 microns) means that it can not filter out bacteria.

• It is hard to pump, and gives a low output rate.

• The tubing is too short, and the plastic parts are fragile.

• Its performance in the field was inconsistent.

• The filter tends to float to the top of the water, and yet it must be submerged in order to work. Thus, one needs three hands (two for the pump, and one to hold the filter underwater).

• The unit doesn't work properly if the water is too shallow to cover the entire filter.

• The foam and filter retain water, so that water leaks all over your pack.

Comments from the Web: None found at the site used.

Summary

The field testers rated this filter as the least favorite of those tested.

8. MSR Waterworks II

This filter was not involved in the *Backpacker* field tests, but we might comment on how it differs from the MSR MiniWorks. The Waterworks II is essentially a larger version of the MiniWorks. It is about twice the price ($125 versus $65), heavier (16.6 oz versus 14.3 oz), requires more pumping force (12 lb versus 8.5 lb), but gives a larger output rate (0.8 liters/minute versus 0.6 liters/minute). It also connects to a 1 liter Nalgene bottle or MSR Dromedary Bag. Whether twice the price is worth a one-third increase in output rate is the issue with this unit.

G. Detailed Comments on Purifiers

As we did above for filters, we will now give some details concerning some of the purifiers, and remarks in the form of "positives", "negatives", and "web site comments."

1. PUR Voyageur

This purifier (see sketch) has a design very similar to that of the PUR Hiker. Its weight, output rate, and strokes per liter are very much the same. The main difference is the filter element, which contains Tritek iodinated resin in the Voyageur. It has no carbon in the element, but the Stop Top carbon filter can be attached. It is remarkably easy to pump (1.6 lb force versus 8.0 lb force for the Hiker).

2. PUR Scout and PUR Explorer

These purifiers have quite a different design from that of the Hiker and Voyageur (see sketch). The pumps have T-shaped handles, rather than the palm-shaped handle found on the Hiker and Voyageur. The Explorer is, in many ways, just a larger version of the Scout. The Explorer weighs twice as much (24.8 oz versus 12.0 oz), costs about twice as much ($130 versus $70), and has a 40% higher output (1.4 liters/minute versus 1.0 liters/minute). It is three times harder to pump (5.0 lb force versus 1.6 lb force); however, 5.0 lb force is still low compared to many other devices. Both units contain iodinated resin, and include Stop Top carbon filters.

So what are the differences, other than size? The primary one appears to be that the Explorer has a brush inside the filter element, which scrubs the filter surface when the pump handle is moved up and down. That is, it is "self cleaning." In addition, the Explorer pump is double-action; that is, it pumps water out on both the upstroke and downstroke.

A few available web site comments generally praise both units, but make mention of a valve on the Scout which stuck, rendering the unit unusable, and the fact that a valve on the Explorer (possibly also on the Scout) can not be cleaned. Another comment relates to the hoses which enter and exit the bottom of the Scout, such that they get pinched when the unit is placed on the ground. Apparently, this has been changed—the hoses now come out of the sides, near the bottom, as on the Explorer.

The PUR Voyageur and Scout also have the 126 square-inch pleated "Anti-Clog" filter which was mentioned in describing the PUR Hiker unit. An accessory called the PUR StopTop Carbon Cartridge (2 oz, $15, carbon refill $6) can be connected to the PUR Explorer, Scout, and Hiker models to remove unpleasant taste and organic chemicals.

The SweetWater Guardian Plus is simply the SweetWater Guardian filter unit with an iodine resin cartridge called the Viral Guard ($25) attached to it. Indeed, this resin cartridge can be connected to the outlet of any water filter to turn the unit into a purifier. A unit called the SweetWater Global Express ($90) is a combination of the SweetWater Guardian Plus connected to a flexible one-liter bottle that flattens as it empties, and a connection kit that allows you to pump directly into the bottle. The pressure-relief valve has been redesigned to give a steady discharge when the filter starts to clog, instead of a spray.

H. Costs per Unit Volume of Water Produced

A comparison of nearly all available filters and purifiers in terms of costs per gallon of water produced, based both on the initial cost of the device and also on the cost of replacement elements, is shown in Table 4.6.

Table 4.6 **Filtration Costs per Gallon for Filters and Purifiers**

Model	Life (gallons)	New Unit Cost	Cost $/gal	Element Cost	Subsequent $/gal
PUR Pioneer	12	$ 30	$2.50	$ 4	$0.33
PUR Explorer	100	$130	$1.30	$ 45	$0.45
PUR Scout	100	$ 80	$0.80	$ 40	$0.40
PUR Voyageur	100	$ 70	$0.70	$ 35	$0.35
First Need	100	$ 70	$0.70	$ 33	$0.33
SW Guardian	200	$ 60	$0.30	$ 20	$0.10
SW Guardian Plus	200	$ 80	$0.40	$ 35	$0.18
SW WalkAbout	125	$ 35	$0.28	$ 13	$0.10
PUR Hiker	200	$ 55	$0.28	$ 25	$0.13
Katadyn Mini	1000	$140	$0.14	$ 60	$0.06
Basic Designs	500	$ 20	$0.04	$ 14	$0.03
Katadyn Wat. Fil.	15000	$250	$0.02	$165	$0.01
Katadyn Combi	14000	$185	$0.01	$ 90	$0.01
MSR MiniWorks	NR	$ 65	-	$ 50	-
MSR WaterWorks II	NR	$125	-	$ 65	-

NR = No reported values.

No lifetime data have been found for the MSR units, so costs could not be computed. Under "element cost," if there are two elements (e.g., an iodine resin element and a carbon element) it is assumed that both are replaced, and the cost shown is the total cost.

One can see an obvious disadvantage of many filters and purifiers relative to chemical disinfectants—their costs per unit of water treated are generally pretty high, based on the initial unit price. With replacement of the filter element after the initial element lifetime, subsequent costs are much more reasonable, but still much larger than the costs of using disinfectants such as iodine solution, iodine tincture, and iodine tablets.

A study has been reported that employed three water sources in Washington State, which varied considerably in terms of suspended matter. For five different filters, the amounts of these waters which could be passed through before the filter elements needed to be replaced were determined. The results, cited in Table 4.7, were determined by the writer by measurement of bars on a bar-chart, and thus may be slightly uncertain. The White River water was the "dirtiest," and Lake Washington water the "cleanest."

Table 4.7 **Liters Pumped Before the Need for Element Replacement**[a]

Water:	Green Lake	Lake Washington	White River
PUR Explorer	40	60	7
SweetWater Guardian	20	65	25
MSR WaterWorks	112	118	21
First Need Deluxe	30	64	12
Katadyn Pocket Filter	550	1915	320

[a] From the Internet at http://www.campmor.com/water filters

The values for the Katadyn unit, in particular, are gross estimates, because the bar chart was difficult to interpret for these high values. It is assumed that the Katadyn filter was cleaned as necessary whenever it became clogged, and thus the values quoted are for the ultimate element lifetime after perhaps 100 cleanings.

In this study, the costs per liter of water shown in Table 4.8 were estimated. Based on these figures, it was deduced by the writer that the costs of the units were taken to be the costs of the replacement elements, which were found to be: PUR Explorer, $45; SweetWater Guardian, $20; MSR WaterWorks, $30; First Need Deluxe, $30; and Katadyn Pocket Filter, $145. (The table in the report gives the MSR costs as 0.55, 0.27, and 0.60, respectively, but these numbers do not appear to be consistent with the bar chart values of liters pumped through the MSR unit.)

**Table 4.8 Costs of Waters Produced by Filters
in the Washington Study ($/liter)**

Water:	Green Lake	Lake Washington	White River
PUR Explorer	1.12	0.75	6.43
SweetWater Guardian	1.00	0.31	0.80
MSR WaterWorks	0.27	0.25	1.43
First Need Deluxe	1.00	0.47	2.50
Katadyn Pocket Filter	0.07	0.02	0.12

One can see that, for the dirtiest water (White River), the water costs are generally quite high. For the cleanest water (Lake Washington), the costs are much better, but are still fairly high for the PUR unit. Costs per gallon would be about 3.8 times the costs shown per liter, so almost all of the values exceed $1.00/gallon ($0.26/liter).

I. Squeeze-Bottle Type Filters and Purifiers

A relatively new purifier called the PentaPure Sport Water Purification System (WTC/ECOMASTER Corp., Plymouth, MN), "Oasis" model, consists of a plastic 18 oz water bottle identical to those that bicycle riders attach to their bikes. The screw-on top has a bent-over sipping tube coming out of its center. One unscrews the top, adds water, and sips clean water out of the tube, while squeezing the bottle. The tube is connected to a three-stage cylindrical cartridge containing, in sequence: (1) a 1 micron filter, PentaPure iodinated resin, and granular activated carbon. The bottle, stated to be good for 200 refills, costs $35, with replacement cartridges selling for $25. It weighs 7 oz, empty. One anecdotal report says that getting reasonable flow requires pretty strong squeezing and sipping. Also, it is difficult to fill the bottle from a shallow water source.

A very similar product marketed by a company called SafeWater Anywhere consists of a squeezable plastic bottle with a straight drinking tube connected to the top. Just inside the bottle, connected to the tube, is a filter assembly consisting of a 25 micron prefilter and a 2 micron proprietary filter, which is said to remove petroleum by-products, trace metals (lead, mercury, cadmium, aluminum), radioactive materials such as radon 222, volatile organic chemicals, pesticides, and insecticides. Thus, the primary component of the 2 micron filter must be activated carbon, in a porous matrix form. A half-liter sized bottle is $35, and a 1 liter size is $40. The unit doesn't kill viruses. To do this, the manufacturer says to add iodine to the water (be sure to wait for a while afterwards), which will be adsorbed by the activated carbon as the water flows out.

J. Some Tips on Using Filters and Purifiers

When using filters/purifiers, a few simple practices can greatly reduce clogging problems.

1. Use the Cleanest Water Source Possible

If taking water from a river or stream, remember that pools—and especially their edges—are cleaner than faster moving current areas (turbulent water stirs up sediment). The end of the intake hose should be kept well away from the bottom. Try to pick a spot where the bottom is relatively free of fine sand, mud, debris, and algae-covered rocks. Use a filter/purifier having a float collar on the intake hose, and place the intake hose carefully.

Despite these precautions, if the water appears to have significant suspended matter, one should scoop it up in a pot, let the water settle for an hour or two (or even overnight, if possible), and then filter the clear water.

2. Backwashing

If your filter/purifier allows this operation (not all do), backwash (or "backflush") the filter when the output rate starts to decrease. That is, make the water flow through the unit in the reverse of its normal direction. This can help lift deposited material from the surface of the filter element and push it out of the unit.

Consult the directions for your unit. Backflushing normally just involves detaching the intake tube and attaching it to the unit's outlet, then pumping for a while.

After backflushing, remember that the outlet side of the unit is filled with whatever your flushing water happens to be. If this water is impure, pump the unit in the normal fashion and discard the first part of it (e.g., pint or half liter). The best water to use for backflushing is clean water previously collected during normal operation.

3. Scrubbing Ceramic Filter Elements

When the output from a unit having a ceramic filter element starts to drop, or the required pumping force becomes unduly large, the filter surface needs to be scraped to remove deposited gunk. Most ceramic units come with a suitable brush. If not, use a stiff toothbrush.

When scrubbing a ceramic filter element, make sure that the material scrubbed off does not fall into the outlet hose, collection bottles, etc. The layer being scrubbed off is likely to be full of pathogens.

4. Don't Mix Your Inlet and Outlet Hoses

Remember that your inlet hose has been in contact with "dirty" water and the outlet hose in contact with "clean" water. Try not to wrap them up together—keep them separate if possible.

While the amount of dangerous material in the few drops of water remaining in a well-drained inlet hose probably is not enough to cause harm, good practice suggests that the inlet hose be handled carefully and kept separate.

5. Long-Term Storage

When storing a filter between trips, the filter should be sanitized by running some chlorinated water (one capful of bleach/quart) through it. It should then be pumped dry or drained. To be extra careful, store the filter in a refrigerator to inhibit microorganism growth (but not a freezer, as some filter elements, particularly ceramic ones, can crack if frozen with residual water in them). However, refrigeration is probably unnecessary if sanitation with chlorine has been performed.

5

Iodine, Iodine Compounds, and Chlorine Compounds

Iodine or chlorine, or compounds containing them, can be effective in killing all types of pathogens—protozoa, bacteria, and viruses, if added in sufficient concentration and if enough time is allowed. *Cryptosporidium* cysts, however, appear to be an exception to this. Generally, bacteria and viruses are killed much more quickly than protozoa—indeed, *Giardia* can survive several hours under certain conditions (cold water, low disinfectant dose). Iodine, and compounds of iodine and chlorine, are cheap and light-weight. A 1-ounce bottle of iodine crystals or tincture of iodine could last some people many years.

An article in *Backpacker* magazine (K. Hostetter, "The Water Filter Field Test," December, 1996, pp. 62-70, 112-115) makes a remark about the use of iodine by the National Outdoor Leadership School (NOLS) and by Outward Bound (OB): "Historically, NOLS and OB participants treat their water with iodine." The article quotes a Mr. Earl Thompsen, equipment manager for an OB school in Texas, as stating: "It's simple, cheap, light, fool-proof, and hassle-free." Mr. Thompson does go on to say that for students allergic to iodine, and when *Cryptosporidium* is of concern, filters have been considered.

Table 5.1 lists available iodine and chlorine products.

Table 5.1   **Available Iodine and Chlorine Formulations**

Name	Active Chemical	Recommended Dose per Quart	Total ppm Concentration	pH
Polar Pure	Iodine crystals 99.5%	1-7 capfuls of saturated pure solution	2.4 ppm per capful	6.1
Coghlan's EGDWT	Tetraglycine hydroperiodide, 16.7%	1 tablet	8	5.6
Potable Aqua DWGT	Tetraglycine hydroperiodide, 16.7%	1 tablet	8	5.6
2% iodine	Iodine	0.4 mL	8	6.5
Halazone	p-dichloro-sulfamoyl benzoic acid, 2.97%	5 tablets	5.4	6.7
Chlorine bleach	sodium hypo-chlorite, 5.25%	0.2 mL	5.0	7.1

The full names, and manufacturers, are: Polar Pure Water Disinfectant (Polar Equipment, Saratoga, CA); Coghlan's Emergency Germicidal Drinking Water Tablets (Coghlan's Ltd., Winnipeg, Canada); Potable Aqua Drinking Water Germicidal Tablets (Wisconsin Pharmacal, Jackson, WI); and Halazone (Abbott Laboratories, North Chicago, IL).

The 2% iodine is equivalent to any standard tincture of iodine, and the chlorine bleach is equivalent to any standard 5.25% sodium hypochlorite bleach.

A. Iodine Crystals

Iodine is a solid halogen having a bluish-black color, and a metallic lustre. It has the unusual property of being able to change from the solid state to a vapor directly, without passing through the liquid state. It will slowly vaporize if left in an unsealed container.

One can add water to iodine crystals in a small glass bottle or vial having a leak-proof hard plastic (e.g., Bakelite) cap, as is the case with the Polar Pure product. Plastic bottles such as polyethylene or polypropylene stain heavily with iodine, and can deteriorate and leak; thus, they are not recommended. One merely lets the iodine dissolve to form a "saturated" solution containing an iodine concentration of about 300 ppm (parts per million by weight, that is, the weight of iodine per million parts of water) or 300 mg/L (for water solutions, ppm and mg/L concentrations are essentially identical, since 1 liter of water is 1,000,000 mg). The saturated solution concentration is independent of the container size; therefore, any size bottle or vial will do. One then pours a given amount of this solution (e.g., 1 fluid ounce, or about 30 mL) into, say, a quart of water, and lets the iodine work for about 30 minutes. One should immediately add more water to the iodine container in order to generate more saturated solution.

It is important to avoid ingesting the iodine crystals themselves, as iodine is moderately toxic (the lethal dose for humans is usually given as 2-4 grams, but fatalities have occurred with less). Thus, when pouring the saturated solution, one must be careful to avoid any carry-over of the crystals. Additionally, some people (5% or less of the population) are allergic to elemental iodine (an allergic reaction to eating shrimp is a tip-off to this) and so individuals should use iodine sparingly until they are sure that no allergic reaction will occur.

The Polar Pure product was examined by the writer. It was found to contain 7.55 grams of solid iodine in the form of small spheres of fairly uniform size. While this amount of iodine is greater than the lethal dose for humans, the bottle has a cylindrical insert lining the neck and extending down to where the bottle diameter increases. As one pours the iodine-saturated water out of the bottle, the iodine spheres fall under the collar and are very effectively retained inside the bottle. The Polar Pure bottle, filled to the top with water and capped, was found to weigh 147 grams, or 5.2 oz.

The side of the bottle has a series of dots, one of which turns green at the prevailing temperature. Various doses in capfuls are noted on the bottle next to the dots. The capful-versus-temperature scale is designed to deliver an iodine dose of 4 ppm into one's water. The cap volume was measured and found to be 7.5 mL. Assuming an iodine concentration of 300 mg/liter (300 ppm), which is exactly correct for a temperature of 71 °F, each capful would deliver 2.25 mg of iodine, giving a concentration of 2.25 ppm if poured into a one-liter bottle, or 2.38 ppm in a quart. This agrees with the figure of 2.4 ppm given in Table 5.1.

At room temperature, the dot which was green indicated that 1.5 capfuls should be used. Thus, the ppm iodine in a quart with 1.5 capfuls would be 3.57 ppm. This may be too low for some situations (cold and/or turbid water).

The instructions that accompany Polar Pure say that, after using some of the iodine-saturated water, one should add fresh water and wait for one hour. After one hour, the solution is said to be ready for use again. The writer did the following test: distilled water was added to the initially water-free bottle, and a sample was withdrawn at different times, its color intensity measured in a spectrophotometer, and the sample then put back into the bottle (each measurement took only about 15 seconds). A final sample was taken after two days to get a definite "saturation" value for the color. Figure 5.1 shows the iodine concentration versus time, plotted as percent of the final saturation value. After one hour, the iodine concentration was only 30% of the final value. Assuming the final value to be 300 ppm, the value at one hour was only 90 ppm. Thus, using the iodine solution after one hour of initially filling the bottle would give a very unsatisfactory iodine dose. What these results mean is that, after the <u>first</u> filling of the bottle with water, one should wait a fairly long time for saturation to be achieved. It is recommended that one do this at home, a day or more prior to the trip on which it is to be used.

Figure 5.1   **Iodine dissolution, with and without shaking**

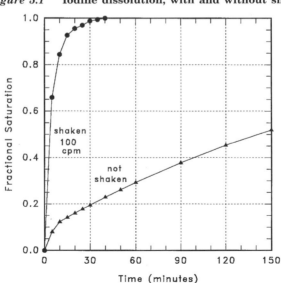

Figure 5.1 also shows the saturation test repeated with constant vigorous shaking of the bottle at 100 cycles per minute (cpm) in a shaking device. The approach to saturation is tremendously enhanced by the shaking—in 10 minutes, the solution is about 84% saturated. With no shaking, a layer of nearly saturated solution "sits" at the bottom of the bottle, surrounding the iodine crystals. This greatly inhibits further dissolution. In contrast, with shaking, the nearly-saturated solution surrounding the iodine crystals is dispersed throughout the rest of the bottle, making the solution next to the crystals far from saturation.

Thus, if one were to have the Polar Pure bottle in one's backpack while one is hiking, the jostling action imparted to the bottle should greatly speed up the saturation process (this situation was not simulated). While the response would not be as good as with vigorous shaking, it would be far better than for zero movement of the bottle. These results also imply that, if one plans to re-saturate the solution at camp, with the bottle sitting idly on the ground, one should shake it periodically—if one desires to accelerate the saturation process.

After the initial filling of the bottle, and attainment of saturation, one only pours out a few capfuls at a time. Since a capful is 7.5 mL and the bottle's capacity is about 55 mL, pouring off 3 capfuls, for example, means that one uses 41% of the saturated solution, leaving 59% of the saturated solution behind. When one adds fresh water back, the new iodine concentration will be only 177 ppm (59% saturation). The new concentration versus time curve is the same as shown in Figure 5.1, except with a starting point of 59% saturation at zero time. Since this figure gives a time of 185 minutes at 59% saturation, we merely add one hour to that, to get 245 minutes, and read the corresponding iodine concentration, about 207 ppm. It is clear that a one-hour wait is not enough to approach anywhere near saturation. Of course, the fewer the number of capfuls used, the higher will be the iodine level remaining in the bottle, and thus the higher will be the iodine level after one hour.

Table 5.2 shows a calculation of the iodine levels after one hour, for different numbers of capfuls used.

**Table 5.2 Estimation of Polar Pure Iodine Levels
One-Hour After Refilling the Bottle**

Capfuls Used	Iodine Upon Refilling (ppm)	Iodine One-Hour Later (ppm)
0.5	280	284
1.0	259	268
1.5	239	253
2.0	218	237
2.5	198	221
3.0	177	207
3.5	157	191
4.0	136	176

These computed results are based on 300 ppm as the saturation value, and the time dependence shown in Figure 5.1 for the unshaken bottle. These computed estimates show that, if one uses several capfuls, one must either wait much longer than one hour for the solution to approach saturation again, or use even more capfuls than indicated, since after one hour the iodine content will be much less than saturation. For example, after using 2.0 capfuls and waiting one hour, the iodine solution is at about 237 ppm, so if one wishes to again use a nominal 2.0 capful dose, one should increase it to 2.0(300/237) = 2.5 capfuls. This, of course, makes the situation even worse an hour later, since the iodine level will fall even further with the next addition of fresh water. The easiest approach to avoiding this "downward spiral" is to try to let the solution saturate whenever possible, such as overnight.

At 7.55 grams iodine per bottle (7550 mg) and a dose of 8 mg/liter, a bottle of Polar Pure could treat 944 liters of water. At a price of $9.00, this would be $0.0095 per liter, that is, 0.95 cents per liter. On the same basis (8 ppm iodine), this is almost identical to the cost of using tincture of iodine (see Table 5.5), and about 10% of the cost of Potable Aqua tablets.

One research study (Jarroll, E. L., Jr., Bingham, A. K., and Meyer, E. A., "Inability of an Iodination Method to Destroy Completely *Giardia* Cysts in Cold Water," West. J. Med. 132, 567, 1980) tested the ability of iodine to kill cysts of *Giardia* in both "clear" water (pH 7.1) and "cloudy" water (pH 7.3) at 3 °C (37 °F) and 20 °C (68 °F) at iodine dose levels of roughly 3-7 ppm. The cysts were essentially completely killed in 15 minutes (stated as 99.8[+] percent inactivation) at 20 °C in both kinds of waters with iodine doses of 3.0-3.3 ppm. Figure 5.2 shows the results obtained at 3 °C. The 4.58 ppm iodine dose with cloudy water is inadequate for only a 45 minute contact time. However, the 6.87 ppm iodine dose with clear water was fairly effective for 30 and 45 minutes contact time, and might well give an almost total kill if one were to wait 60 minutes.

Figure 5.2   Jarroll's results for Giardia with Iodine at 3 °C.

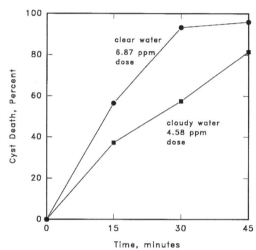

The authors imply that iodine was a failure, but their results are not unexpected. As mentioned above, at 20 °C, even 3.0-3.3 ppm iodine was effective in 15 minutes, in both waters. This is a good performance. At quite low temperatures, it is common wisdom that the iodine dose and time of waiting should both be increased, e.g. to 8 ppm iodine for clear water, somewhat more than 8 ppm iodine for cloudy water (e.g., 12-16 ppm), and 60 minutes. Unfortunately, Jarroll did not investigate what an 8 ppm iodine dose would do at 60 minutes time for either type of water. However, there is a clear lesson to be learned from the data: temperature has a very strong effect on the disinfection process, and cold waters require substantially higher iodine doses and times.

The same research group also studied the effectiveness of other chemical disinfectants on *Giardia* cyst inactivation in clear and cloudy waters, at 3 °C and 20 °C (Jarroll, E. L., Jr., Bingham, A. K., and Meyer, E. A., "*Giardia* Cyst Destruction: Effectiveness of Six Small-Quantity Water

Disinfection Methods", Am. J. Trop. Med. Hyg. 29(1), 8, 1980). The amounts of each disinfectant used, per liter of water, and the contact times were: Halazone, 5 tablets, 30 minutes; bleach (5.25%), 0.1 mL (2.5 ppm) in clear water and 0.2 mL (5 ppm) in cloudy water, 30 minutes; Globaline (which is essentially the same as Potable Aqua tablets), 1 tablet in clear water and 2 tablets in cloudy water, 20 minutes; Coghlan's Emergency Drinking Water Tablets (CEDWGT), 2 tablets and 20 minutes for cloudy water and for clear water at 3 °C, and 1 tablet and 10 minutes for clear water at 20 °C; iodine tincture, 0.5 mL (10 ppm) for cloudy water and 0.25 mL (5 ppm) for clear water, 30 minutes; iodine saturated solution, 26 mL (about 7.8 ppm) and 20 minutes for cloudy water and for clear water at 3 °C, and 13 mL (about 3.9 ppm) and 15 minutes for clear water at 20 °C.

For both clear and cloudy water at 20 °C, the doses used were effective (99.8+ percent kill) for every disinfectant form. The greater doses used at 3 °C for cloudy water were also completely effective (99.8+ percent kill), except for the saturated iodine, which gave a 77.3% kill. Increasing the time from 20 to 30 minutes would probably correct this deficiency. At 3 °C for clear water, Halazone and CEDGWT gave 99.8+ percent kills, but the Globaline, saturated iodine solution, bleach, and iodine tincture treatments gave less than complete kills—97.5, 96.5, 91.7, and 74.6%, respectively.

These deficiencies could be corrected easily by (1) increasing the contact time for saturated iodine from 20 to 30 minutes, (2) increasing the iodine tincture dose from 0.25 mL (5 ppm) to 0.5 mL (10 ppm) , (3) using 2 tablets rather than 1, or 30 minutes contact rather than 20 minutes, for the Globaline, and (4) increasing the bleach dose from 0.1 mL (2.5 ppm) to 0.2 mL (5 ppm). Indeed, the disinfectant doses used in this study at 3 °C for clear water seem too low, since temperature has a large effect on the rate of disinfection.

B. Tetraglycine Hydroperiodide Tablets

One can buy tablets containing iodine in the form of a compound named tetraglycine hydroperiodide which, when added to water, give about 8 ppm iodine in the water. These tablets were developed back in the early 1950s by researchers at Harvard University, under research grants from the U. S. Army. The tablets contain the buffer disodium dihydrogen pyrophosphate ($Na_2H_2P_2O_7$) to lower the water pH (one tablet will lower the pH to 5.6, as shown in Table 5.1) and increase the fraction of iodine present as I_2 rather than its less effective HOI form (see Chapter 6). This formulation came to be known as Globaline and was marketed with that name for many years. More recently, this compound has been marketed under the two names shown in Table 5.1.

These tablets usually incorporate a lubricant such as talc (to make the tablets fall out of the tablet press easily), and a swelling agent such as bentonite (to promote disintegration in water), and some kind of filler (such as sodium chloride), as well as the buffer.

These tablets are reputed to have only a moderate shelf life; however, studies have shown that tablets in sealed amber-glass bottles, stored in a 100% humidity environment at 130 ºF, lost only 21% of their active iodine over a period of 24 weeks (of course, with the bottles well-sealed, the high humidity was not in direct contact with the tablets). When the tablets were exposed to room air in open dishes at 60 ºC (140 ºF), there was an iodine loss of 40% over 7 days, and when they were exposed in open dishes to 100% humidity at room temperature, they lost 33% of their iodine content in 4 days (in comparison, Halazone tablets lost 75% of their chlorine in only 2 days of such exposure).

Other tests showed that well-protected tablets (in bottles with wax-sealed caps) dissolved rapidly, in one minute or less, but that those in ordinary capped bottles tended to harden with age in warm and humid environments, causing slow dissolution. One advantage of these tablets over crystalline iodine and Halazone is their much superior rate of dissolution, if reasonably "fresh" tablets are used. Studies have shown that Globaline tablets dissolve in less than one minute, whereas Halazone tablets dissolve in about 7.5 minutes.

The tablets are said to turn color from gray to yellow as they deteriorate, so a visual inspection for color may be useful as a gross indicator of potency. The usual dose is one tablet per quart of clear water, and two tablets per quart of cloudy water. Doses should be increased if the water is quite cold. A contact time of 30 minutes minimum is recommended (more for cold water).

The writer weighed a group of tablets on an analytical balance and found the average weight per tablet to be about 113 mg. The bottle label states that "each tablet contains 6.68% titratable iodine." Thus, the iodine equivalent in each tablet is 0.0668(113) = 7.55 mg, and so one tablet per liter of water would give 7.55 ppm iodine. The concentration in a quart (0.946 liter) would be 8.0 ppm.

From the fact that the total iodine in a bottle of Potable Aqua is about 0.38 g, one advantage of these tablets over iodine crystals is that one would be carrying far less than a lethal dose of iodine.

At 50 tablets per bottle of these tablets, a bottle would treat 25-50 liters or quarts of water, at a dose of 1-2 tablets per liter or quart. At a cost of about $5.00 per bottle, this is 10-20 cents per liter or quart.

C. Tincture of Iodine

A quantity of 100 mL of "2%" tincture of iodine is made by dissolving 2 g of pure iodine crystals with 2.4 g of sodium iodide in 50 mL of pure ethyl alcohol (the kind of alcohol in beer, wine, and distilled spirits), and adding sufficient pure water to bring the volume to 100 mL. Thus, the tincture is 2% iodine on a weight per volume basis (2 g/100 mL, or 20 mg/mL).

The writer has measured drop volumes produced by various types of droppers, and found that they varied between about 0.025 mL and 0.050 mL, depending on the shape of the end of the dropper and the hole size. A typical average value was about 0.04 mL. For a drop of 0.04 mL volume, each drop of 2% iodine tincture would deliver 0.04 mL x 20 mg/mL = 0.8 mg iodine. Thus, 5 drops of the tincture would give 4 ppm iodine in a liter of water, and 10 drops of the tincture would give 8 ppm iodine in a liter of water.

Proper doses in terms of numbers of drops of tincture and ppm iodine will be discussed in some detail in Chapter 6. For now, let us just say that the iodine dose should be increased if the water is cloudy and/or cold. If one can decide the ppm iodine level that is appropriate for the water involved, one merely multiplies the ppm value by 5/4 to determine the number of drops of tincture to use.

The writer purchased a bottle of iodine tincture, labeled "1 fluid ounce, 30 mL" (a fluid ounce is actually 29.6 mL, but the pharmaceutical industry rounds this off to 30 mL) for $0.72 (this includes 6% tax) at a Wal-Mart in Laramie, Wyoming, in June of 1997. This bottle thus contained 600 mg of iodine. At a dose of 8 ppm, or 8 mg/liter, this could treat 75 liters (79 quarts, or about 20 gallons) of water. The cost would be 0.96 cents per liter (0.91 cents/quart, 3.6 cents/gallon), or about 10% of the cost of Potable Aqua. Clearly, this is dirt cheap. The bottle is small, very light (about 1 oz), and would last for about two weeks if one were to treat 3 gallons/day of water. If one is not allergic to iodine, and is willing to wait the required time for disinfection, the use of tincture of iodine clearly has tremendous advantages (it is the method that the writer has used for many years).

It should be noted that a typical 1-ounce tincture of iodine bottle does not come with a dropper type of dispensing head; rather, the bottle opening is fairly large (e.g., 3/8 inch in diameter). To use the tincture on a dropwise basis, one needs to buy a drop-type dispensing bottle and put the tincture in that. The writer has used a small eye-drop bottle for this. The drop-forming top of such a bottle can be snapped off, allowing the eye-drop solution to be poured out, and the bottle rinsed with tap water. The iodine tincture is then poured in, and the snap-on top replaced. It is obviously extremely important that one then label the bottle clearly with the word "iodine," using a permanent marker, so that there is no chance of it being used in one's eyes.

Iodine does have the disadvantage just mentioned: one must wait for 30 minutes to as much as several hours for it to work completely (a detailed discussion of this will be given later). And, it results in the water having the "band-aid" taste of iodine. This taste can be neutralized by adding some pure ascorbic acid (vitamin C) or a drink mix containing some ascorbic acid. The neutralization occurs instantaneously after the vitamin C dissolves (thus, one should add the vitamin C after waiting for the iodine to do its job). Vitamin C is very inexpensive, and easy to carry.

D. Chlorine Products

The chlorine products listed in Table 5.1 are much less effective than iodine, and should not be used. Halazone has a relatively short shelf life (5-6 months) and loses potency rapidly (75% loss in 48 hours) when exposed to air, heat, and/or moisture. In addition, it tastes terrible. Reports indicate that vitamin C can be used, as in the case of iodine, to neutralize the taste of chlorine. Chlorine is not as effective as iodine at pH values above about 7.5, and tends to bind easily to nitrogenous pollutants, thereby reducing the level of the "free" or "residual" chlorine, the only form that is effective. Additionally, the reaction of chlorine with some industrial wastes and decayed organic matter can produce carcinogens such as chloramines and trihalomethanes (e.g., chloroform, bromoform).

The recommended dose is 5 tablets/quart for clear water and 10 tablets/quart for muddy water. For water at ambient temperatures, 30 minutes contact time (minimum) is recommended. Since a bottle contains 50 tablets, one bottle is sufficient only for 5-10 quarts of water. At a cost of $6.50 per bottle (estimate), this amounts to about $0.65 to $1.30 per quart, rather expensive as compared to any of the iodine methods.

The writer weighed a group of Halazone tablets on an analytical balance and determined an average weight of 138 mg per tablet. The bottle label states that the active ingredient, p-(dichlorosulfamoyl) benzoic acid, is 2.97% of the weight, or 4.1 mg per tablet. The bottle label says to use 5 tablets per quart. Simple calculations show that this would give a chlorine dose of about 5.4 ppm. The label also states: "Allow water to stand for 30 minutes after tablets have dissolved before drinking."

Liquid household chlorine bleach can be used to disinfect water, if sodium hypochlorite (NaOCl) is its sole active ingredient. At 5.25% sodium hypochlorite, each mL contains 25.0 mg chlorine. To achieve 5 ppm chlorine in a liter of water, one needs to use 0.2 mL of the bleach.

Granular and powdered bleaches contain other chemicals, and are poisonous. Five drops of bleach (0.2 mL total, or 5 ppm, at 0.04 mL/ drop) per liter of clear water and 30 minutes contact time are usually recommended, if the water is at ambient temperature. Liquid bleach loses 50% of its power over one year of storage. It is generally not powerful enough to kill protozoa such as *Giardia*.

There are other chlorine products, not listed in Table 5.1, which have come to the attention of the writer. These appear not to be widely marketed. Indeed, only one (AquaCure) has been seen by the writer, and in only one store. We will now describe them.

First is a chlorine product called AquaCure, made in South Africa and marketed in the USA by SafeSport Manufacturing Co., Denver, CO. It is sold as a package of 30 tablets ($7.95, or 26.5 cents/tablet) with a composition of 2.5% sodium dichloro-s-triazinetrione and 97.5% inert ingredients. Actually, the inert ingredient part contains alum as a flocculent. The dose for 77 °F (25 °C) water is one tablet (600 mg) per liter. This provides "1.4% available chlorine", which would be (1.4/100) x 600 = 8.4 mg chlorine, that is, 8.4 ppm chlorine in a liter. For 41 °F (5°C) water, two tablets per liter is the recommended dose.

One problem with this product is that, if sufficient settling of the alum is not allowed, one ends up drinking significant alum, which tastes horrible. One example: The writer took a hiking trip with a friend, who was allergic to iodine and bought this product, and put a tablet in a liter of water. The bottle, being in his day-pack, was thus subjected to a back-and-forth motion, which prevented settling of the alum. We stopped to drink some water, and I asked to drink some of his, to see what it was like. The water was white and cloudy, and had a strong astringent taste.

Another chlorine product is Sierra Water Purifier (4-in-1 Water Co., Santa Fe, NM) which comes in a two-part kit. One part is a small bottle of calcium hypochlorite, $Ca(OCl)_2$, crystals, which one adds to the water. The other bottle contains 30% hydrogen peroxide (which is extremely corrosive and can burn skin), which one then adds after the hypochlorite has done its job of disinfection. The peroxide reacts with the hypochlorite to form soluble calcium chloride, an odorless and essentially tasteless compound. Excess peroxide bubbles off as oxygen. The doses recommended are 50 mg of the hypochlorite crystals and 6 drops of the peroxide, per gallon.

Yet another product to be used in combination with hypochlorite compounds is called Cl-Out. It consists of a small wand with brush-like zinc/copper alloy bristles. One places this in previously chlorinated water and stirs. The zinc catalyzes an electrochemical reaction which reduces the hypochlorite to chloride, which has no color or odor, and little taste. It will also reduce iodine to iodide, which also has no color or odor, and little taste. The zinc is not used up, and thus the product lasts forever. This method works only with relatively small amounts of water at a time, as one must stir the water enough so that all of it eventually comes in contact with the bristle. However, ascorbic acid (vitamin C) will cause the same chemical reduction of hypochlorite or iodine, and works instantaneously once it is mixed into the water. Thus, ascorbic acid is easier to use than both Cl-Out or the peroxide part of the Sierra Water Purifier.

Having discussed the costs of various chemical disinfectants, we can easily compare their costs. Table 5.5 shows a comparison based on achieving 8 ppm with Polar Pure iodine crystals, 8 ppm with 2% iodine tincture, the use of one tablet/quart of Potable Aqua or Sierra Water Purifier, and the use of 5 tablets/quart of Halazone.

Table 5.5 **Costs per Liter or Quart for Various Chemical Disinfectants**

Product	Cost (cents)	Volume
Tincture of Iodine	0.96	liter
Polar Pure Iodine Crystals	0.95	liter
Potable Aqua Tablets	10	quart
Sierra Water Purifier	27	liter
Halazone Tablets	33	quart

The writer has been unable to locate Halazone in any store or catalog, and has only a bottle (100 tablets) purchased about 10 years ago for $6.49. Assuming no change in price, even if Halazone is still marketed, this would amount to 6.5 cents per tablet, or about 33 cents for 5 tablets/quart, the recommended dose. Clearly, tincture of iodine and iodine crystals are much cheaper than the tablet formulations.

E. Comparative Tests of the Methods Using Giardia Cysts

A very illuminating study of the effectiveness of various water purification methods has been carried out by researchers, using *Giardia lamblia* cysts harvested from gerbils who had been infected with cultured trophozoites. One liter volumes of water containing 30,000 cysts were passed through various filters, and the entire outlet water streams were examined for the presence of cysts. A First Need filter and a Katadyn Pocket filter removed 100% of the cysts, while a filter named H_2OK (H_2OK Portable Drinking Water Treatment Unit Model No. 6, Better Living Laboratories, Memphis, TN) was 90% effective. A filter named the Pocket Purifier (Calco Ltd., Rosemont, IL) did not remove a statistically significant amount of the cysts. The units were each tested three times, and the results averaged.

The efficacies of chemical treatments were tested using a clear water and a turbid water, both at 10°C. The chemical treatment products listed in Table 5.1 were evaluated. The doses used were stated to be "according to manufacturers' instructions" (presumably those in Table 5.1), but no details are given (e.g., as to how many capfuls of Polar Pure solution were used). In each test, a 50 mL water sample was inoculated with cysts, mixed, incubated at 10°C, and 10 mL portions were withdrawn at 0, 30 minutes, and 8 hours time. Cyst viability was determined on each portion by a staining procedure, which distinguished live and dead cysts by their different colors. Table 5.6 shows the percent survival of the cysts for each treatment method, for the iodine compounds and for the chlorine compounds, respectively. The "control" results are for samples handled with the standard procedure but with no chemical added. All numbers were estimated by careful measurement of bar-chart results.

It is clear that the chlorine compounds were very ineffective, even after 8 hours contact, the only exception being for Halazone with clear water. Even in this case, 9% (or 2700) of the cysts survived, which is much more than enough to cause infection.

The iodine compounds were also surprisingly ineffective at 30 minutes contact time, although they were far superior to the chlorine compounds. However, at 8 hours, all of the iodine compounds inactivated the cysts completely. It is unfortunate that a contact time somewhere between the extremes of 30 minutes and 8 hours was not also studied, e.g. 2 hours. It may very well be that 2 hours would also produce a 100% kill with the iodine compounds.

Table 5.6 **Effectiveness of Different Chemical Treatments in Killing** *Giardia lamblia* **Cysts (figures are percent survival)**[a]

	Polar Pure	CEGDWT	Potable Aqua	2% Iodine	Halazone	Bleach	Control
Clear water, 30 min	22.6	15.1	7.8	29.6	61.0	61.0	58.2
Turbid water, 30 min	38.4	27.8	27.8	48.2	67.3	81.0	76.4
Clear water, 8 hr	0	0	0	0	9.1	27.9	61.6
Turbid water, 8 hr	0	0	0	0	21.8	46.2	66.5

[a] Ongerth, J. E., et al., "Backcountry Water Treatment to Prevent Giardiasis," Am. J. Pub. Health 79, 1633, 1989.

It must be noted that the temperature involved was 10 °C. It can be estimated (using the disinfection "model" discussed in Chapter 6) that the death rate at 20 °C would be about 3 times faster. Thus, if one used a contact time of 30 minutes with water at 20 °C, a 100% kill might easily be achieved.

In another aspect of the same study, so-called contact disinfection units were evaluated. One was the H_2OK filter and another was the Pocket Purifier. These devices claim to inactivate cysts, as well as removing them by filtration (we saw earlier that they filter poorly). A third unit, named the Water Purifier (Water Technologies Corp., Ann Arbor, MI) does not claim to be a filter, but does claim to be a contact disinfection device. It too was tested. Again, cyst inactivation was determined by the staining method, with the results being the average of three replications. The Pocket Purifier, H_2OK, and Water Purifier devices inactivated 2%, 5%, and 15% of the cysts in low turbidity water—clearly inadequate.

Inactivation of cysts by heating was also assessed by incubating samples of distilled water plus added cysts for 10 minutes at temperatures ranging from 10 to 70 °C. Whereas the percent inactivation at 40 °C was only about 10%, the percent inactivations at 50 and 60 °C were about 90% and 97%, respectively. At 70 °C, inactivation was 100%. Thus, heating water to 70 °C (158 °F) for 10 minutes appears to kill *Giardia* completely.

Because this study was carried out with real cysts, its results are undeniable and extremely valuable. The most important conclusions are: (1) chlorine compounds are ineffective for killing *Giardia*, even over an 8 hour period, (2) iodine compounds are effective, if the contact time is sufficient (what is "sufficient" depends strongly on the water temperature, as we will see in Chapter 6), (3) the "contact" disinfection devices worked poorly, and (4) heating to 70 °C for 10 minutes is 100% effective for cysts, at least for clean waters.

6

Details About How Iodine Works and Doses to Use

The first few pages of this chapter are somewhat technical. Readers who are not familiar with chemistry and mathematics might wish to go directly to Section D.

A. Solubility of Iodine in Water

The solubility of iodine in water depends on temperature as shown in Figure 6.1. Table 6.1 gives precise values at specific temperatures.

Table 6.1 **Saturation Iodine Concentrations in Water Versus Temperature**

Temperature (Fahrenheit)	Iodine Concentration (ppm)	Temperature (Celsius)	Iodine Concentration (ppm)
35	148	0	140
40	162	5	166
45	179	10	200
50	200	15	242
55	223	20	285
60	246	25	334
65	270	30	385
70	295	35	444
75	322		
80	350		
85	378		
90	410		

Figure 6.1   Solubility of Iodine in water

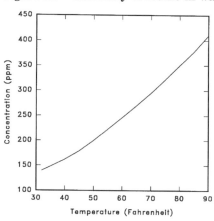

At 20°C (68°F) the concentration of iodine crystals in equilibrium with water (that is, the saturation concentration) is about 285 ppm, or 285 mg/L. To give an iodine concentration of 8 ppm in a quart of water, 27.3 mL of this saturated solution will be required (i.e., slightly less than a fluid ounce, 29.6 mL).

B. Iodine Dissociation

When iodine is dissolved in water, it undergoes a "hydrolysis" reaction:
$$I_2 + H_2O \rightleftarrows HOI + I^- + H^+$$
The HOI (hypoiodous acid) which is formed by the reaction of I_2 with H_2O may subsequently ionize to form the hypoiodide ion OI^- according to:
$$HOI \rightleftarrows OI^- + H^+$$
Table 6.2 shows, for an I_2 dose of 8 ppm, how much of the iodine is in the various forms at 68 °F (20 °C), as a function of pH.

Table 6.2   **Effect of pH on the Forms of Dissolved Iodine at 68 °F for a 8 ppm Initial Iodine Dose**

pH	Percent of Iodine in Each Form		
	I_2	HOI	OI⁻
5	96.8	3.2	0
6	90.2	9.8	0
7	72.3	27.7	0
8	37.2	62.8	0
9	8.0	92.0	0
10	0.6	99.1	0.3

Since I_2 and HOI are both effective as disinfectants (although HOI kills cysts at only about 30-50% of the speed of I_2), whereas OI⁻ is ineffective, there is no loss of disinfection potential until the pH is well above pH 10, a situation that would occur rarely in practice.

pH is defined as: pH = - log[H⁺], where [H⁺] denotes the concentration of H⁺ ions in moles/liter. Water can dissociate according to:

$$H_2O \rightleftarrows H^+ + OH^-$$

Pure distilled water has [H⁺] = 1 x 10⁻⁷ moles/liter, and thus its pH = 7. A pH of 6, however, means that [H⁺] = 1 x 10⁻⁶ moles/liter, and therefore the concentration of H⁺ is 10 times higher than at pH 7. Indeed, each unit on the pH scale represents a 10-fold change in the concentration of H⁺.

C. A Model for Iodine Disinfection

The simplest model for inactivation of an organism such as a cyst, which we will call X, by a chemical agent, which we will call C, is represented by the reaction $nC + X \rightleftarrows C_nX$, where it is assumed that "n" molecules of C are needed to "kill" the cyst. X represents the living cyst, and C_nX represents the dead, or inactivated, cyst. The rate of change in X with time can be represented by: Change of X with time = $-k[X][C]^n$

The square brackets indicate concentrations, e.g., the amounts of X and of C per unit volume. The minus sign indicates that X decreases with time. The quantity k is a "rate constant" which characterizes the speed of the inactivation process. k depends on the type of organism, temperature, and pH. The rate law can be "integrated" mathematically to give a relation between the viable cyst concentration at any time (X) and the viable cyst concentration at the start of the process (X_0), that is, at the time the chemical is added: $X/X_0 = \exp(-kC^nt)$

The right side of the above equation, "$\exp(-kC^nt)$" means the quantity "e" (2.718) raised to the power $-kC^nt$, where t = time, and the equation merely states that the concentration of living cysts decreases exponentially with time. The whole problem in using these kinds of equations to describe the rate of killing any particular organism is to determine exactly what the value of k is for that organism, at the prevailing temperature and pH of the water. This is no simple task.

Chura, in a 1986 MS Thesis titled "Small Scale Disinfection of Drinking Water to Prevent Giardiasis" (University of Wyoming, Laramie) modeled the inactivation of *Giardia lamblia* cysts using the equation just described, with n = 1.4 (a value obtained by others), and an equation relating the rate constant k to temperature. Details may be found in his thesis.

D. Recommended Iodine Doses

Chura's model was evaluated for various iodine doses, temperatures, and pH values. The times necessary for 99.9% cyst destruction (this is a commonly-used criterion of organism inactivation, since these models predict infinite time for 100% inactivation) were determined.

Figure 6.2 shows the times required for disinfection as a function of the iodine dose and temperature. The times rise rapidly as the iodine doses fall below 6-8 ppm. The effect of temperature is not strong for iodine doses above the 6-8 ppm range, but is very pronounced for lower iodine doses. This temperature effect can be seen better in Figure 6.3, where we see that the times for 4 ppm iodine are indeed much more dependent on temperature than those for 8 and 12 ppm iodine. To assess the effect of pH, disinfection times were computed as a function of

Figure 6.2 Disinfection Time versus Iodine Dose at Different Temperatures

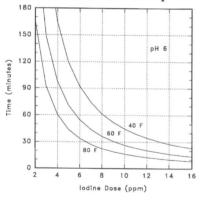

Figure 6.3 Disinfection Time versus Temperature for Various Iodine Doses

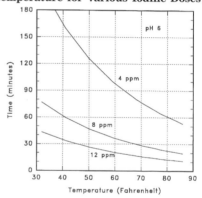

Figure 6.4 Disinfection Time versus pH for 8 ppm Iodine

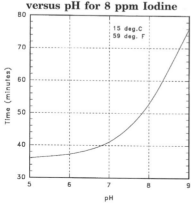

Figure 6.5 Iodine Dose versus Temperature for 30 and 60 Minute Waiting Time

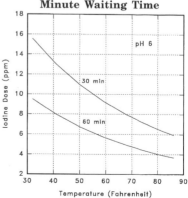

pH for 59 °F (15 °C) and an 8 ppm iodine dose. Figure 6.4 shows that below a pH of 7, the times are not strongly dependent on pH, but above pH 8, they are. The reason for this is simple—at pH 7 and below, the iodine is predominantly in the form of I_2, but as the pH rises it converts more and more to HOI, which is less effective.

A more valuable way of using Chura's model is to ask the question: "If I were willing to wait 60 minutes, or only 30 minutes, what iodine dose would be required for clear waters at different temperatures?" Figure 6.5 answers this question. For a 99.9% kill in 30 minutes, the iodine dose ranges from almost 16 ppm down to 6 ppm, over the temperature range 32-86 °F. For a 99.9% kill in 60 minutes, the iodine doses range from almost 10 ppm to about 4 ppm over the temperature interval 32-86°F.

Figure 6.5 therefore represents recommended iodine doses based on a waiting time approach. These doses should be increased slightly if the pH is 7, and even more if it is higher.

Typical cold free-flowing stream and river waters usually are in the pH range of 6-7. Thus, for a temperature of 41 °F (5 °C), the iodine dose should be about 8 ppm for 60 minutes waiting time, or 13 ppm for 30 minutes waiting time. On the other hand, mountain lake waters are usually in the pH range of 5-7 (acid rain causes pH values near 5, sometimes even a bit less). Lakes in the Eastern USA (e.g., the Adirondacks of New York State) are often of pH 4.5-5, and mountain lakes in the Rocky Mountain region are typically pH 5-6. For warm pH 5-7 lake water at 68 °F (20 °C), the iodine dose need be only 5 ppm for a 60 minute wait and only 8 ppm for a 30 minute wait.

Waters of pH 9 and above are not usually found in the wilderness, except for prairie lakes in areas where the soil is alkaline due to significant soda ash (Na_2CO_3) and sodium bicarbonate ($NaHCO_3$) content. From Figure 6.4, one can estimate that disinfection times at pH 9 are roughly double those at pH 5-7. Thus, it is recommended that one use the Figure 6.5 dose for 30 minutes and simply wait 60 minutes.

Table 6.3 gives values of the iodine doses in ppm and the number of drops of iodine tincture, based on Figure 6.5.

Table 6.3 **Recommended Iodine Doses and Iodine Tincture Drops**[a,b]

Clear water, 30 minutes contact time

	0°C(32°F)	5°C(41°F)	10°C(50°F)	15°C(59°F)	20°C(68°F)
ppm	16	13	11	9	8
drops	20	17	14	12	10

Clear water, 60 minutes contact time

	0°C(32°F)	5°C(41°F)	10°C(50°F)	15°C(59°F)	20°C(68°F)
ppm	10	8	7	6	5
drops	13	10	9	8	7

[a] For "dirty" water, double all numbers. [b] For pH values greater than 6, increase the doses and drops by the following percentages: pH 7, 10%; pH 8, 50%; pH 9, 100%.

All numbers have been rounded up to the next higher integer, and the drop volume has been assumed to be 0.04 mL (0.8 mg iodine/drop). The doubling of the doses for dirty waters is because they are likely to have substances which may chemically bind iodine, leaving less "residual" for disinfection. Some substances, such as tannins released by decaying leaves, can actually react with iodine, producing iodide ions (I^-), which are ineffective for disinfection. The dirtier the water, the greater is the likelihood that binding or reaction of the iodine will occur, and thus the greater the iodine dose must be. The amount of dose to give dirty waters is a difficult "judgement call," and the writer assumes no responsibility for the "doubling" guideline.

The numbers of drops of iodine tincture cited in Table 6.3 need to be adjusted if one suspects that one's drop volume is not 0.04 mL. Some medical bottle type droppers give as little as 0.025 mL per drop. It is up to each reader to try to assess the drop volume of the dropper bottle being used. This can be done using a common small measuring spoon (preferably metal rather than plastic, since iodine stains plastic easily) and the fact that 1 tsp (teaspoon) equals 4.93 mL. Thus, 1/8th tsp = 0.62 mL. If one counts the drops needed to fill a 1/8th tsp spoon and divides this into 0.62, one can determine the mL/drop fairly accurately. Suppose one does this and finds that 18 drops are needed. Then the volume per drop is 0.62/18 = 0.034 mL. Each drop represents 0.034 mL x 20 mg/mL = 0.68 mg iodine. If one wants 8 ppm iodine in a liter of water (i.e., 8 mg iodine), then one needs to use 8/0.68 = 11.8 drops (rounded up to 12). Alternately, note from Table 6.3 that 8 ppm requires 10 drops of 0.04 mL volume. One then writes 10 x 0.04/0.034 = 11.8 drops, the same result.

If one is using saturated iodine solution (e.g., Polar Pure), recall that such a solution is about 300 ppm, and thus each 4 ppm of desired concentration requires about 30 mL, or 1 fluid ounce, as computed earlier. When using tetraglycine hydroperiodide tablets, recall that one tablet in a quart gives anywhere from 4.5 to 7.7 ppm, depending on which value is correct. It is recommended that one use the lower figure of 4.5 ppm for one tablet/quart.

7

Other Considerations
When Using Disinfectants

This chapter considers a variety of aspects related to the use of iodinated or chlorinated water.

A. Adding Flavoring Agents to Iodinated Water

A research article published in 1968 noted that military personnel often added soft drink mixes (e.g., Kool-Aid) to water supplies treated with iodine tablets, in order to mask the unpleasant taste of the iodine. The researchers discovered that iodine reacts with ascorbic acid (vitamin C), an additive commonly found in drink mixes. Thus, they strongly recommended that water flavoring agents containing vitamin C not be added until the iodine has done its job.

The reaction between iodine and ascorbic acid is: I_2 + ascorbic acid $\rightarrow$ dehydroascorbic acid + 2 I^-. Since the iodide ion, I^-, has very little germicidal power, this reaction destroys the effectiveness of I_2.

Cooney and Chura (Cooney, D. O. and Chura, J. P., "Flavoring Agents and the Disinfection of Water", J. Envir. Eng. <u>116</u>, 642, 1990) did a study in which the speed of this reaction was determined. They found that the reaction was complete in less than 0.03 seconds. They also studied the reaction of iodine with tannic acid, a component of ice tea mixes.

Again, a conversion of I_2 to I^- occurred, this time in less than 0.18 seconds. Tests of other flavoring agent components such as citric acid, sodium saccharin, and aspartame (NutraSweet) showed that no reaction occurs with these, and so drink mixes having these are safe to add at any time (however, it is always a good policy to wait until disinfection is complete, just to be certain).

In the same study, Cooney and Chura prepared an 8 ppm solution of iodine in water, and heated it on a laboratory hot plate. They measured the temperature at regular intervals, and took small samples of the solution, which were later analyzed for iodine concentration. The idea was to determine if heating water, after the iodine had inactivated organisms, would drive off the iodine and thus improve the water's taste.

Figure 7.1   **Evolution of Iodine as Water is Brought to a Boil**

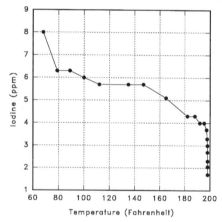

Figure 7.1 shows the results. It is clear that the evolution of iodine is quite slow—by the time the water has reached about 150 °F, the iodine concentration is still almost 6 ppm. Since heating water to this range of temperature will itself kill most organisms, heating the water to drive off iodine defeats the purpose of the iodine.

Hence, it is recommended that, to get rid of the taste of iodine in water, one should simply add some vitamin C (preferably crush the tablet first, so that it will dissolve faster) or just add a drink mix containing vitamin C. Indeed one can now buy "two part" kits for disinfection (Potable Aqua Plus, $8), one bottle being tetraglycine hydroperiodide tablets and the other bottle containing vitamin C tablets.

B. The Safety of Iodine

Iodine is a component of the thyroid hormone thyroxin, which helps regulate growth, development, and metabolic rate. If one ingests excessive amounts of iodine, a condition know as goiter can occur, in which the thyroid gland is enlarged and thyroid activity is lowered. Thus, it is important to assess whether the amount of iodine ingested through iodination of wilderness waters poses any potential threat.

A 1968 research study described the use of iodine to disinfect water in two water systems supplying three Florida prisons having a total population of about 700 people. Disinfection was very effective. During the first 17 months of the 19 month test period, the iodine dose was 1.0 ppm; in the last 2 months of the period, it was 0.3 ppm. Careful examinations were conducted on 70 subjects. Three indices of thyroid function were evaluated. No change in the physical characteristics of the subjects' thyroid glands were found, and there was no evidence of "any detrimental effect on general health or thyroid function." No allergic reactions to the iodine were observed. This study indicates that chronic exposure to iodine at a level of 1.0 ppm is harmless; however, it did not prove anything about exposure to significantly higher levels, even if relatively short-lived.

However, in a military test conducted in 1953, soldiers were given much higher iodine doses than in the 1968 study—average doses of 12 mg/day for 16 weeks and then 19.2 mg/day for 10 weeks. There was "no evidence of weight loss, vision impairment, cardiovascular damage, altered thyroid activity, anemia, bone marrow depression, or renal [kidney] irritation among the subjects." Since 12-19.2 mg/day represents the ingestion of 3-5 liters per day of water having 4 ppm (4 mg/liter) iodine, this test duplicated typical, and perhaps somewhat more than typical, daily ingestions of iodine by a backpacker.

C. Chlorine Dissociation

Readers may be interested to know what happens when chlorine dissolves in water, for comparison to what occurs with iodine, and so we give a brief summary here. The dissociations which take place are of exactly the same type as those with iodine.

When chlorine is dissolved in water, it, like iodine, undergoes the following hydrolysis reaction to form hypochlorous acid, HOCl.

$$Cl_2 + H_2O \rightleftharpoons HOCl + Cl^- + H^+$$

The HOCl which is formed may subsequently ionize to form the hypochlorite ion (OCl⁻) according to

$$HOCl \rightleftharpoons OCl^- + H^+$$

Table 7.1 shows, for an Cl_2 dose of 8.0 ppm, how much of the iodine is in the various forms at 68°F (20 °C), as a function of pH.

Table 7.1   **Effect of pH on the Forms of Dissolved Chlorine at 68 °F for a 8 ppm Initial Chlorine Dose**

pH	Percent of Chlorine in Each Form		
	Cl_2	HOCl	OCl⁻
1	0.91	99.09	0
2	0.09	99.91	0
3	0.02	99.98	0
4	0	99.97	0.03
5	0	99.74	0.26
6	0	97.45	2.55
7	0	79.27	20.73
8	0	27.67	72.33
9	0	3.68	96.32
10	0	0.38	99.62

A dose of 8 ppm was used rather than the 4 ppm dose we employed with iodine, because chlorine is less effective than iodine and therefore its recommended dosage level is higher than that of iodine. We show numbers for unrealistically low pH values (1-4) only to illustrate that dissolved Cl_2 hydrolyzes rapidly to HOCl above about pH 3. Thus HOCl is the predominant form of chlorine over the pH range of 4 to somewhat above 7.

Since HOCl is a very effective disinfectant, whereas OCl⁻ is ineffective, the disinfection power of chlorine declines rapidly above a pH of 7 (i.e., at alkaline pH values). This is the reason why Halazone tablets contain a buffer whose role is to keep the pH on the acidic side, although Table 5.1 indicates that the pH level they produce, 6.7, is only slightly less than neutral.

The objection to chlorine as a disinfectant, as compared to iodine, is not just that it is poorly effective in alkaline waters, but also that it tends to react more strongly with organic matter in water, leaving less residual or free HOCl, and causing the creation of chloramines and/or trihalomethanes, which are carcinogenic. In addition, chlorine is simply less effective, ppm for ppm, than iodine. Finally, the tablet form of chlorine (Halazone) has a poor shelf life, and deteriorates rapidly upon being exposed to air, light, and moisture. For all of these reasons, chlorine and chlorine compounds are not recommended.

8

Summary And Recommendations

We will assume, for purposes of this chapter, that we are indeed dealing with true wilderness or backcountry waters containing no industrial, agricultural, or sewage discharges. We will also assume for the moment that there are no cattle or sheep upstream of the point of water collection, although in many parts of the USA, especially in the West, there is considerable grazing by cattle and sheep on US Forest and BLM (Bureau of Land Management) lands which are otherwise wilderness in nature. We will refer to waters which are free of impact from any human activity and domesticated animal grazing as "pristine."

Let us look at all of the water treatment options, and decide which ones can be eliminated from further consideration. First, with respect to chemical disinfectants, we have seen that chlorine compounds are generally less effective, less stable, and much more expensive than iodine or iodine compounds. They also react more strongly with organic matter, thereby reducing the free chlorine residual, and they also react with nitrogenous compounds to form carcinogenic chloramines. There is little reason to choose any chlorine compound over iodine or iodine compounds.

Now, focusing on the iodine options, it is clear that 2% tincture of iodine has many advantages over iodine crystals and tetraglycine hydroperiodide tablets. Iodine tincture is cheap, lightweight, indefinitely stable, and instantly soluble in water. A bottle of iodine crystals is larger than a small "dropper" type bottle of iodine tincture, and the rate of dissolution of the crystals is slow. The chance of ingesting a toxic dose of crystals is greater than ingesting a toxic dose of iodine tincture (one can easily see and count drops of iodine tincture, whereas carry-over of iodine crystals from a bottle of crystals may not be noticed). Thus, of the iodine options, we are left with tincture of iodine as the superior choice.

Tincture of iodine is almost a "magic bullet," with the advantages of effectiveness, very low cost, very light weight, and instant solubility in water. It does entail the disadvantage of one having to wait 30 minutes or so for it to work, but that is ordinarily not a problem—one merely uses two water bottles, and lets one undergo the 30 minutes of disinfection while one uses water from the other, previously disinfected, bottle. Vitamin C, or drink mixes containing vitamin C, can be used to nullify the taste of the iodine, after disinfection is complete.

The one drawback to tincture of iodine is that it does not kill crypto cysts. However, we have seen earlier that waters which can be called pristine typically have crypto concentrations on the order of 0.004-0.2 cysts/liter, and that as many as 5 cysts might be needed for infection. Even assuming that, say, 2 cysts will produce infection (we know that 1 cyst is insufficient) and that the water has 0.2 cysts/liter (the upper end of the data for pristine waters), one would have to drink 10 liters of water, more or less in "one sitting," to run a significant risk of infection. If one's assessment of the water one is drinking is that it is pristine, then tincture of iodine should be sufficient treatment. On the other hand, if the area one is traveling in is suspect, in that it may contain discharges from human activity, contamination by cattle or sheep, or even a significant likelihood of contamination by common carriers of protozoa, such as beavers, then one must be much more conservative. In this case, a filter should be used to remove protozoa and bacteria.

Based on the field tests reported in *Backpacker* magazine, one might logically consider the PUR Hiker, the SweetWater Guardian, and the MSR Miniworks filters—the top three favorites by far in the *Backpacker* field test. These units combine fairly low cost ($55-$65 range), with fairly light weight (11-14 oz range), and relative ease of pumping. To inactivate viruses, one can either add iodine to the filtered water (this again involves waiting for disinfection to occur) or consider using purifiers having iodine resin elements, such as the PUR Voyageur, PUR Scout, or SweetWater Guardian Plus (the PUR Explorer is much heavier than these three, and much more expensive). The SweetWater Guardian and Sweetwater Guardian Plus have activated carbon in their elements, and the PUR units can be combined with the StopTop carbon accessory, in the event that organic chemical contamination is of concern.

The other filters and purifiers which are available all have some serious drawbacks, such as much higher cost, much higher weight, high required pumping forces, and pore sizes which are too large to remove bacteria.

The two squeeze-bottle units described might also be considered, although they are difficult to fill from shallow water sources. The SafeWater Anywhere unit should be combined with the use of iodine drops, since it has no iodine component in its filter and its pore size is 2 microns—far too large to remove bacteria. The PentaPure bottle unit does have iodinated resin, as well as activated carbon, in its three-stage filter. Assuming it is not too hard to squeeze (not tested by the writer), it looks promising. One's logical choices are therefore:

1. Just use tincture of iodine (see Chapter 6 for the appropriate ppm levels, drop numbers, and waiting times) if one believes crypto cysts and organic chemicals are not likely problems.

2. Use a filter which will remove cysts and bacteria, if it is believed that crypto cysts may be a problem, and that viruses are not a problem (since most filters can not be counted on to remove viruses). In general, viruses pose little risk in the USA and Canada, but they are often found in other parts of the world.

3. Use a filter having activated carbon in it, or an attachable carbon cartridge, if it is believed that organic chemicals and crypto cysts are both likely to exist.

4. Most certain of all, use a purifier with an iodine resin element to remove cysts and bacteria, and to inactivate viruses.

5. Most completely of all, select a purifier having activated carbon in it (e.g., the SweetWater Guardian Plus) or an attachable carbon cartridge (e.g., in the case of the PUR Scout and Voyageur units). This will take care of all microorganisms and organic chemicals.

Table 8.1 gives a summary of the removal capabilities of some of the more popular filters and purifiers discussed.

Table 8.1   **Removal Capabilities of Selected Filters and Purifiers**

Cysts Only	Cysts, bacteria	Cysts, bacteria, org. chem.	Cysts, bacteria, viruses	Cysts, bacteria, viruses, org. chem.
Timberline Eagle	First Need Deluxe	MSR Water-works II	PUR Scout	PUR Scout*
Basic Designs 2050	Katadyn Pocket Filter	SweetWater Guardian	PUR Explorer	PUR Explorer*
	Katadyn Minifilter	PUR Hiker	PUR Voyageur	PUR Voyageur*
	MSR Mini-works			SweetWater Guardian Plus

* With StopTop accessory attached.

Again, we mention that the First Need Deluxe has been placed in the category of non-removal of viruses, as a conservative measure. The manufacturer claims it removes viruses, but there is no proof of this at present.

Glossary

Depth filter: Any filter made of materials such as glass fibers packed to a significant depth. As the water flows in a random circuitous path through the bed of packed fibers, particles and microorganisms are trapped at different points. Because the effective pore sizes of passages in the bed vary widely in size, trapping is a statistical process. Only when a particle or microorganism tries to pass through a sufficiently small passage does physical entrapment occur.

Ceramic filter element: A ceramic filter has well-defined pores, and so any particles or microorganisms larger than that pore size are strained out at the surface, rather than deep within the ceramic. This is why ceramic elements can be rejuvenated by scraping their outer surfaces.

Activated carbon: Activated carbons are made from various starting materials, such as wood, coal, and coconut shells. The material is heated up in the absence of air to drive off volatile matter and produce a char (or charcoal) which is mostly carbon (90% or more). The char is then "activated" by exposing it to steam at high temperatures. The steam reacts with the carbon and etches it, creating a network of fine pores inside the char. The total surface area of the pores is huge—typically 1000-1500 square meters per gram (about 11,000-16,000 square feet per gram). This surface is capable of adsorbing (binding) organic chemicals, thereby removing them from contaminated water. Once the adsorption capacity of the activated carbon is exhausted, it must be replaced.

Resin: The resins used in water purifiers are small, hard beads made from polymers such as polystyrene, to which iodine groups (I_3, I_5) are attached by chemical bonds. These beads have the property of releasing small amounts of the iodine when water is passed through a bed of the beads.

Organic chemicals: Chemicals comprised primarily of carbon, hydrogen, and oxygen, sometimes with minor amounts of other atoms such as sulfur, nitrogen, and halogens (e.g., chlorine). Molecules comprising living matter such as vegetation are usually organic. In contrast, inorganic chemicals are comprised mainly of metallic atoms and their compounds. Most mineral matter ($NaCl$, $CaCO_3$, $FeSO_4$, etc.) is inorganic.

Figure Legends